Seascape

Seascape

Notes from a Changing Coastline

By Matthew Yeomans

2025

Copyright © Matthew Yeomans, 2025

All rights reserved. No part of this book may be reproduced in any material form (including photocopying or storing it in any medium by electronic means and whether or not transiently or incidentally to some other use of this publication) without the written permission of the copyright owner. Applications for the copyright owner's written permission to reproduce any part of this publication should be addressed to Calon, University Registry, King Edward VII Avenue, Cardiff CF10 3NS.

www.uwp.co.uk

British Library Cataloguing-in-Publication Data
A catalogue record for this book is available from the British Library.

ISBN: 978-1-83760-030-4

The right of Matthew Yeomans to be identified as author of this work has been asserted in accordance with sections 77 and 79 of the Copyright, Designs and Patents Act 1988.

For GPSR enquiries please contact: Easy Access System Europe Oü, 16879218. Mustamäe tee 50, 10621, Tallinn, Estonia. gpsr.requests@easproject.com

Cover artwork by Neil Gower
Cover design by Andy Ward
Typeset by Agnes Graves
Printed and bound in Great Britain by Bell & Bain Ltd, Glasgow

To Jowa, Dylan and Zelda

Contents

	Introduction	ix
1.	Taming the Levels	1
2.	Turning Back the Tide	18
3.	Castles in the Sand	34
4.	Copperopolis	52
5.	Putting Faith in Science	66
6.	Below the Landsker Line	85
7.	The Sustainable Sea	99
8.	Stories of the Sea	113
9.	Farewell to Fairbourne	127
10.	The Town That was Built on a Beach	142
11.	Whose Home is This Anyway?	157
12.	Two Bridges over Troubled Waters	172
13.	The Wreck that Inspired the Shipping Forecast	186
14.	The Winds of Change	199
	Selected Bibliography	216
	Acknowledgements	218

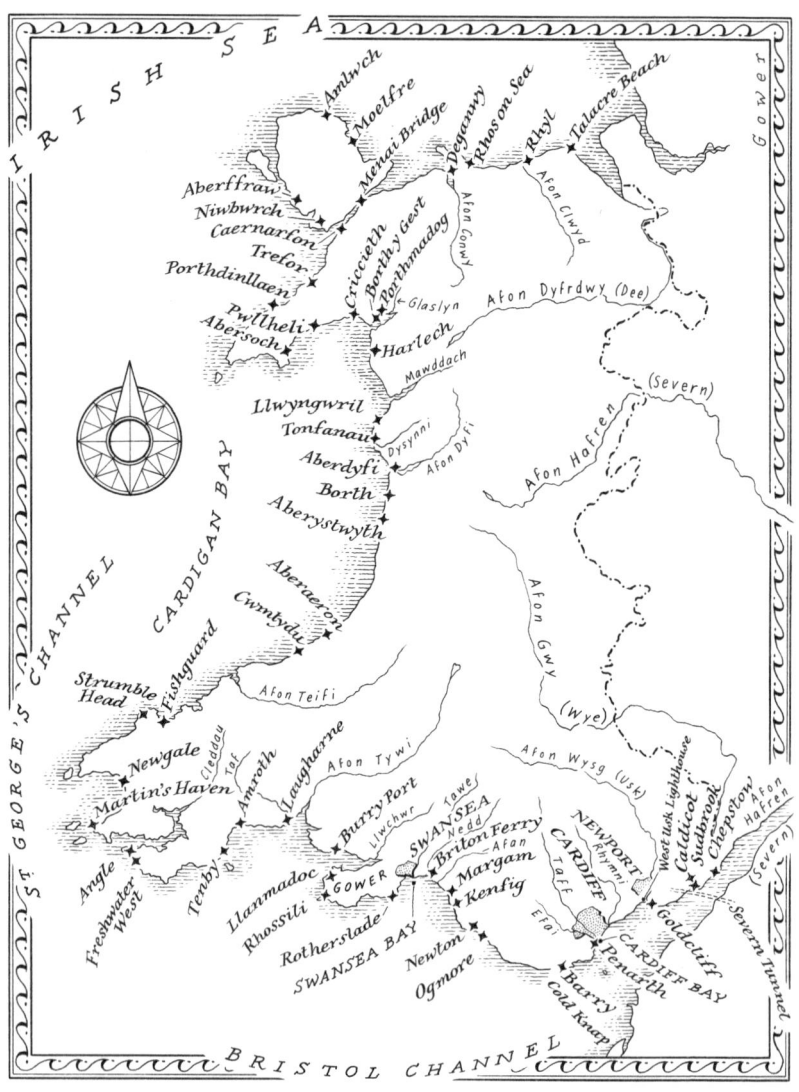

Introduction

What do you think about when you see the sea? How does it make you feel? Are you excited? Scared? In awe of its power and beauty? Maybe you don't think that much about it at all. If that's the case, you probably live a long way inland.

For the millions of people living on or close by the coast, the sea is a dominant presence. It creates and delivers most of our weather, it shapes and changes the physical coastline we call home. It has been and continues to be instrumental in our economy. And it shapes our culture and our sense of identity.

For many of us, it engenders a crucial sense of balance and well-being. I can sit for hours looking at the sea and the waves. Calm days are good but, often, stormy days are better. I love losing myself in its many moods and rhythms, marvelling at its power, its energy and yes, sometimes, its destructiveness.

Not too long ago, I was doing just that on a late autumn afternoon at Dunraven Bay on the south Wales coast. It was near high tide and the waves overpowered the flat, shelf-like rocks that run down to the sandy beach. A few wetsuit-clad surfers tested their skills in the murky grey water. I was here looking for a bit of inspiration – counsel from the sea if you will – for a new project.

Seascape

I had recently finished a book that had taken me wandering through the woodlands of Wales. I tramped through parts of the country I had never discovered before and the experience opened my eyes to the power of walking and the magical scenery of mountainous Wales. I had also come to understand the importance of reconnecting with nature for my own peace of mind. I wanted to keep my Welsh adventure going and so I gravitated back to the sea – my natural happy place.

Part of my inspiration for a new walking project was the Wales Coast Path – the official walking trail that covers 870 miles (1,400 km) of the Welsh coastline. It had just celebrated its ten-year anniversary and I was intrigued at the prospect of walking parts of it – losing myself in the beauty and beaches of Gower, Pembrokeshire, Ceredigion, Ynys Môn (Anglesey) – and all points in between.

I also wanted to learn more about how the sea has shaped our lives through history; how Wales's rich mythology, literature and culture has been influenced by this fluid force, and how our sense of identity has been formed and continues to evolve in relationship to the changing and growing role that the sea plays in our lives.

I had brought a book with me to the beach – a weighty tome titled *Wales and the Sea: 10,000 Years of Welsh Maritime History*. Something written in its introduction made me think hard about how much we have forgotten about the sea:

> Wales has always been a seafaring and sea-dependent nation, its history connected with water. Today, this is less obvious, now that road and rail systems and life- and workstyles of people in Wales have largely turned away from the sea. Once it was sailors who brought back home tales of foreign lands and different cultures.

Introduction

It was an important point. Too often there exists a perception of Wales as an insular, inward-looking nation. Worse, that we are a people obsessed with, and living in the shadow of, our direct land-based neighbour, England. Had that always been the case? Surely not a century ago when Wales was exporting coal all over the world from its coastal ports. But what about before then? And what about Wales's sense of its own position in the world and its sense of self-worth since its industrial decline?

There were so many questions to answer and I was eager to immerse myself in Welsh sea culture. But the more I researched different coastal areas to walk, one theme kept emerging – the looming impact of climate change on the people, wildlife and coastline. Back in 2020 the National Strategy for Flood and Coastal Erosion Risk Management in Wales had warned that as climate change brings rising sea levels and more intense storms, Wales will face 'tough decisions on how to defend low-lying coastal areas and fluvial floodplains, in particular along estuaries and in our steep-sided valleys.'

Some estimates project that sea level rise, alone, could account for seventy per cent of flooding in Wales by 2080. More than 245,000 properties are at risk from river or coastal floods. Other projections show that large sections of low-lying Welsh coastline could be under the average high tide level in just thirty years.

I have been studying and writing about climate change for nearly two decades. I had written endless times of how, by 2050, two-thirds of the world's population will live in cities – the majority located on the coast. And I've read over and over again how an estimated 800 million people will be vulnerable to at least half a metre rise in sea level and that more than 150 million people currently live on land that will be swamped within thirty years.

Yet, despite having that knowledge, I don't think that I fully appreciated just how short a time frame thirty years really is, or processed how quickly our world might start changing. In many ways, I accepted the realities of climate change in an abstract fashion – impacts that clearly would reshape the world but ones that I had a hard time applying to my daily life. If I was struggling to turn the theoretical into the tangible, how must most of the population be coping?

The problems our society has in comprehending the enormity of climate change are multiple but two clearly stand out. The first is the time delay. People are being asked to contemplate what their world might be like a generation from now. Most of us struggle to plan a year or so ahead.

If it's a challenge to think that far into the future, maybe it's worth thinking back to what you were doing in 1995… exactly thirty years ago? Some readers might be too young to remember but many more will, like me, recall all too well their lives back then.

I was a young journalist in 1995, living in New York and excited by the possibilities of life. In work, I was helping my publication think about how to create a new way of communicating to its readers – it was called a website. I had a mobile phone but no clue about how it would dominate nearly every waking hour of our lives. I had never heard of electric vehicles and if you'd asked me about artificial intelligence, I'd have told you it was just a crazy script idea from the Terminator movies. Back then I was just starting to get interested in environmental and energy issues. But the idea that we might get most of our electricity from harnessing the wind and not from oil and gas? I would have said you were crazy.

Also, if you'd told me then that we would be facing an existential crisis that threatened global society, I'd have had a hard time comprehending how that could be possible or what could be done to make a difference. And yet we were being told about it even then. The scientist James Hansen first warned the US government about man-made climate change in 1988. The world's governments agreed to the formation of the United Nations Framework Convention on Climate Change in 1992. But almost no one paid attention because this concept of climate change seemed so abstract at the time.

Introduction

Since then everyone's awareness of the term climate change has increased immeasurably but not in ways that have made global society take the action needed to prevent its worst consequences, and certainly not enough to make individual people fully grasp the changes that are coming.

That's because the second problem many people – particularly in the developed world – have in understanding climate change is that they haven't experienced it first-hand. Climate change happens to other people – whether it be drought, famine and mass migration from the Horn of Africa, water shortages and social tensions in Syria, or wheat, rice and other staple crop failures that are causing food shortages and price spikes throughout the developing world.

Recently, though, that perception has started to change as more and more people in the developed world have finally started waking up to the direct threat posed by the climate emergency. Wildfires, floods and other extreme weather events ravage the United States and Europe on an increasing and increasingly worrying basis. Average global temperatures are on the rise – making parts of southern Europe and the US unbearably hot during summer months. Consumers in the developed world also are starting to come to terms with specific food shortages and rising costs. And all over the planet, wealthy countries are receiving increasing waves of climate-driven migrants.

Meanwhile, climate change is becoming more intense. According to the International Cryosphere Climate Initiative, since 2021, Antarctic sea ice has hit back-to-back record lows, a winter heatwave melted snow up to 3,000 metres high in the Andes and glaciers in Switzerland have lost ten per cent of their volume. That's before the world reaches 2°C of atmospheric warming. If (but most likely when) that happens, the Arctic Ocean will be ice-free almost every summer and carbon emissions from its thawing permafrost soils will equal those of the European Union today. At the start of 2025, the EU's climate and weather institute, Copernicus, confirmed what many people already suspected: 2024 was the hottest year ever recorded – followed by the year before and the year before that.

Here in Wales, sea level rise and increased flooding – both directly from the sea and from river run-off flowing down from the hills – poses our greatest known threat. Yet, for the most part, we have yet to witness or comprehend fully the changes our climate and the sea will bring – even though the signs are already apparent. As many climate writers and experts have observed, humanity is a bit like the frog that sits contentedly in a pot of slowly boiling water until it perishes.

Sitting by the sea that autumn day, I realised just how much climate change would influence any story I wanted to tell about Wales and its relationship with the sea.

I had recently been reading Jon Gower's book about the Irish Sea, *The Turning Tide*. In it he recounts how, in 1703, Daniel Defoe had compiled eyewitness accounts of the incredible storm that hit the British Isles that year. Defoe used those anecdotes – including two from Wales – to write a book, *The Storm*, which is considered the first modern form of journalism and one of the first attempts to understand the sea and the weather by analysing data and first-person accounts. It got me thinking about Wales's evolving relationship with the sea. Defoe had relied on other people's stories to understand the storm that swept through the British Isles. I had the opportunity to take a different approach. I could both research the past and learn first-hand what is happening along the Welsh coast today.

I decided that I would consider the coast of Wales through the lens of how it (and the lives of those who live there) will be reshaped by the sea in the coming decades. I researched a specific series of walks that could help me make sense of how

Introduction

things might change. The walks would begin in south Wales and head west through Gower and Pembrokeshire to St Davids on the far west flank of Wales. Then I would head north, tracking sections of the coast through Ceredigion, Gwynedd, the Llŷn Peninsula and Ynys Môn. Finally, I would rejoin the mainland and head east again, following the almost linear coastline to Talacre Beach where it meets the Dee Estuary. That would be my end point.

I wouldn't be walking every step of the Coast Path but I would be visiting every coastal region of Wales. In doing so, I hoped I could explore the forces that created the coast we know today, the history of how we humans have shaped the coastline and tried to tame the sea, and the stories and legends passed down through the centuries to make sense of the mysteries of the ocean.

This was also going to be a personal journey partly because so many of my memories and experiences come from spending time on the Welsh coast – from day trips along the south Wales coast and the Gower Peninsula, to getaways further afield in west Wales. Part of my family hails from the seaside town of Tenby and, growing up in the port city of Cardiff, the sea felt like a constant part of my life. I still find myself gravitating back to it when I need to reflect or just decompress from daily life.

As I watched the waves crashing against the rocks and dragging stones back down the beach at Dunraven Bay, it also dawned on me that this would be a journey through Wales's energy and engineering past and future. So much of the climate emergency has been sparked by our fossil fuel addiction. The Welsh coast, through its coal ports and industry, fuelled the addiction long before oil and gas came to dominate our energy use. Yet it was also the Welsh coast that established copper as a core component of the industrial age – now one of the key elements needed for a renewable energy future. Most important, those waves I was looking at, and the winds that accompanied them, represent the purest form of energy on earth.

Seascape

Ultimately this journey explores how we can successfully adapt to life on the coast in the climate change age and how we can work with the sea to create a more sustainable life for all of us. Along the way, it considers five fundamental issues all of us will have to consider.

The first addresses how we got into the coastal crisis we now face: how we have tried to tame the sea to shape the coast in ways that suit our industries and lifestyles.

The second considers the decisions we will need to make on what parts of the coast to protect and how we do that in a way that doesn't compound the problems we already face.

The third is both sobering but also very practical. When it comes to those places that we have to surrender to the sea, where do we relocate the people who live there and how do we achieve that in a way that doesn't put pressure on other parts of society?

The fourth issue builds on the third quite literally. How do we plan and construct differently so that we respect, learn from and sometimes mimic nature rather than trying to control it?

Finally, we are entering an era of great upheaval and of mass migration and changing fortunes. How do we shape a more fluid society so that it becomes stronger as it adapts to the growing influence of the sea.

I don't pretend that this book will deliver all the answers to those questions. But by exploring them, and by gaining a better understanding of how the sea has shaped our lives through past centuries, I hope we can rediscover a connection, appreciation and respect for it. And, if we start thinking and planning carefully for a future that will present us with many unknown challenges, perhaps we can also discover many opportunities for a better way of living.

1

Taming the Levels

Walks from Chepstow to Sudbrook and Caldicot to Goldcliff

In 1880 a young man from Wiltshire was invited to visit south Wales. His name was Henry Fleuss and, despite being just twenty-nine years old, he was already the talk of Victorian society because of his new invention: an apparatus that allowed divers to breathe without being connected by tubes to a supply of oxygen.

Fleuss had been summoned by the Great Western Railway engineering team tasked with building a new tunnel under the Afon Hafren (River Severn). Months earlier, disaster had struck when a key section had flooded. Water poured in at a rate of 360,000 gallons per hour and with such ferocity that the entire tunnel was engulfed in just twenty-four hours. This torrent of water was so powerful and unrelenting that the engineers called it The Great Spring.

In the wake of the disaster, a new chief engineer took charge of the tunnel project. His name was Sir John Hawkshaw. He had previously built Charing Cross and Cannon Street railway stations in London and even advised on the Suez Canal. He quickly set about building new permanent pumping shafts to counter the flow of The Great Spring. These succeeded in diverting most of the water away from the tunnel but it still flowed in from a flood door that had been left open when the workers scrambled to escape the breach.

Someone would have to swim more than 1,000 feet (300 metres) through the tunnel, and in complete darkness, to shut the door. Hawkshaw's team first summoned Alexander Lambert, a professional diver whose fearless demeanour earned him the nickname 'pocket Hercules'. But when he descended into the tunnel, his air-hose-connected helmet became tangled around wooden timber beams and he had to turn back. A different approach was required.

That's when the call went out to Fleuss. His new, hoseless, mobile design involved a copper tank contained in a crude backpack filled with compressed oxygen and a material that absorbed carbon dioxide.

The tunnel builders believed that Fleuss would undertake the dive himself but it soon became clear that the young man was a better inventor than he was a swimmer. After three aborted attempts, Fleuss told the engineers that he wouldn't go back down into the cold, dark, flooded tunnel even if they offered him £10,000.

After much persuasion (and a crash course in using the new breathing apparatus) Lambert agreed to head back underground. Over the course of two dives, he succeeded in closing the door. The railway tunnel was back on track.

On 1 December 1886 the first passenger train successfully passed through what we now know as the Severn Tunnel. Fleuss continued to improve on his invention. Today, he is considered the pioneer of modern scuba diving.

The Wales Coast Path starts (or ends, depending on which direction you're coming from) in south Wales at the town of Cas-Gwent (Chepstow), which just happens to be where my old friend Andy Bethell lives.

Andy accompanied me on some of my previous walks through the woodlands of Wales and, despite that experience, he has agreed to join me once again – this time exploring the coast of Wales. My goal is to walk sections of the path between here and Talacre Beach on the far tip of the Dee Estuary. I want to explore Wales's relationship to the sea – specifically, how the sea has shaped our coastline and the lives of Welsh people through history and how we have tried to control and harness the sea for our own purposes.

Andy makes us a coffee and we sit peering out through his kitchen window at the rainy gloom outside. Today's walk is a gentle 'loosener' from the start point of the path to the village of Sudbrook about six miles away – the scene of Alexander Lambert's dramatic dive and where the Severn Tunnel emerges.

We consult the OS Explorer map that I've got on my laptop. It shows a green diamond dotted line as it snakes overland from the banks of the Afon Gwy (Wye) before reaching the Hafren Estuary and heading southwest, hugging the coast. 'Even you will have to work hard to get us lost on a national walking trail that hugs the coast,' says Andy.

It would be very easy to stay indoors, have another cup of coffee and take a literal rain check but we both agree that having decided to start exploring this long-distance walking route, we are unlikely to make much progress if we wait for fine weather.

We lace up our walking boots, pull on waterproof trousers, zip up our jackets and leave the warm house. The sky is slate grey. A soft, persistent drizzle quickly coats our outer layers – the sort of sneaky rain that doesn't feel cold but chills you to the bone.

We wander down the hill from Andy's house, through the grounds of Chepstow's impressive Norman castle towards the Gwy. Two large, engraved stone monoliths mark the start/end point of the Wales Coast Path. Next to them is an oval metal sculpture that was commissioned to celebrate its tenth anniversary. It is meant to symbolise a pebble. More than a few locals feel it looks more like a baked potato wrapped in silver foil.

This giant metal pebble might seem a surreal way of commemorating the nation's coastal walking path but surely no more so than starting the trail in a town located two miles from the actual coast. There is a logistical explanation at least – one of the aims in creating the path was to link up with the Offa's Dyke national walking trail that runs the north/south length of the Welsh and English border. This way Wales can boast a continuous walking path circumnavigating the entire country. If that meant shifting the coast a couple of miles upstream – well it would hardly be the first geographical sleight of hand to befall this border town.

Back in 1536, when King Henry VIII stripped the Marcher lordships – the Norman descendants who had controlled the border lands of Wales and England since the eleventh century – of their power, Chepstow became part of the newly created county of Monmouthshire. Henry placed it under English jurisdiction and, for the next 400 years, arguments raged over whether Chepstow was part of Wales or England. Today, it is firmly integrated back into Cymru – at least geographically.

And, in fact, Chepstow has maritime pedigree as a port. The Norman invaders quickly identified how this section of the steep-sided Gwy valley provided the easiest crossing point for miles around. William the Conqueror commissioned an intimidatory stone castle to be built at Chepstow and its influence grew from there. Over the centuries, the town's docks became an export hub for timber and bark from nearby Wentwood Forest. In the eighteenth century, Chepstow handled a greater quantity of tonnage than the nearby coastal ports of Caerdydd (Cardiff), Abertawe (Swansea) and Casnewydd (Newport) combined.

Despite the rain, we pause for a minute to take in the river and the steep white limestone cliffs that rise above it, imagining just how bustling this port once must have been. A faded historical plaque informs us that we are standing next to the old town slip that once enabled small boats to launch and allowed cargo from larger ships to be landed.

But that's as close as we are going to get to Chepstow's old maritime history because now the Coast Path runs into its first major obstacle – an apartment building located where the trail should continue. Thirty minutes later, after a detour through the streets of Chepstow and a grim industrial estate, we at last get a first real glimpse of the Welsh coastline. The misty drizzle hits our faces at exactly the right angle to achieve maximum annoyance but we can just make out England on the far side. To our left, the old Severn Bridge ascends and descends onto dry land at the small village of Aust. Between the two coasts lies a swirling, churning, muddy mix of sea and river – the highly potent, often treacherous and quite awe-inspiring witches' brew of a waterway that is the Hafren.

The Hafren is the largest estuary on the British west coast; only the Wash on England's east coast is bigger. Its greatest claim to fame is its dramatic tidal range, rising a full fifteen metres twice a day from low to high tide – one of the largest changes in the world after Canada's Bay of Fundy.

The funnel-like geography of the estuary is what drives this tidal range and gives it power. At its widest point, between Cardiff in Wales and Weston-super-Mare in England, the estuary stretches out nine miles. More than twenty-five miles further upstream, between Black Rock and Aust, the crossing distance shrinks to a mere two miles. Tidal power forces the sea up this funnel, rapidly filling the estuary.

Around 130 days a year – depending on a combination of factors including lunar movements, barometric pressure and the amount of freshwater flowing down the Hafren – the sea transmogrifies into a cluster of tidal waves known as the Severn Bore that race upstream for nearly twenty-five miles. During these occasions, an army of surfers, paddleboarders and even some lunatic bodyboarders rush to key launch points along the river to ride the waves which can rise as high as two metres on particularly strong tides. The bore is so powerful that it reverses the freshwater flow of the Hafren – quite a feat given this is the longest river in the UK, its source starting over 220 miles away on the bare slopes of Mynydd Pumlumon high in the Cambrian Mountains of mid-Wales.

We can thank the Norman travel writer Geoffrey of Monmouth for the English name of the river and the estuary. Back in the twelfth century, he wrote *Historia Regum Britanniae* (History of the Kings of Britain). This tome is best known for popularising the legends of King Arthur but it also recounts the tragic story of Princess Hafren who was drowned in a great river by her evil stepmother Gwendolen. Geoffrey refers to Hafren by an English name, Sabre, and recounts how the river came to be known as Sabren. Earlier, when the Romans settled on the banks of the Hafren they referred to it as Sabrina. Through the ages the Roman name mutated into Severn.

We cross a railway track and the enormous span of the cable-stayed M4 road bridge comes into view. It is only then that it dawns on me exactly where we are headed. In front of us lies Black Rock, the site of perhaps one of the earliest known, and certainly the most trafficked, entry points into Wales. It gets its name from the notable protrusion into the estuary of smooth black rocks – what geologists refer to as Mercia mudstone (sadly not a character from *The Flintstones*) which formed in the Triassic period some 230 million years ago.

It is easy to see why humans have been crossing this section of the estuary for many thousands of years. The English side seems within touching distance – especially when the tide is low and the full expanse of Black Rock is exposed. Nevertheless, and despite the giant bridge in front of us, it is still remarkable to contemplate that this unassuming piece of the coast has played such an outsized role in Wales's connection to the outside world.

We know from archaeological digs that people have been traversing the Hafren from Black Rock since Bronze Age times. Mesolithic fish traps have been found in the mudflats offshore and an important Iron Age Silurian hill fort once dominated the cliffs above Sudbrook. It would have been both a trading post and a military stronghold overlooking the low-lying estuary land for miles around – sending a statement to the rival Dobunni clan across the estuary. It also posed a significant challenge for the Roman legions who arrived on the banks of the Hafren around AD 49.

As is so often the case, it was those Romans who really put their stamp on the place. Here, in the shadow of the giant road bridge, a local tourism informational sign describes how: 'The Romans were regular users of the crossing, ferrying legions from Aust.' That makes sense given that they built a road from their stronghold in Cirencester to the village on the opposite side of the estuary. Roman coins have been found in the mud here at Black Rock and archaeologists speculate they could have been thrown into the river to placate the gods and so ensure a safe passage across the turbulent waters.

Not a great deal more progress was made getting people over the Hafren for the next 1,500 years. Even in the twentieth century motorists had to take a car ferry across the estuary. That was until 1966 when, finally, the Severn Bridge opened a few miles upstream from where we were standing.

At the time, the bridge was lauded for the way it opened up new opportunities for a south Wales region that had been in the freefall of industrial decline for decades. Even then, though, that access came at a high price. The bridge had cost £8 million to build. To recoup the investment, the UK government decided that drivers would have to pay a toll to use it.

In truth, people had paid a fee to cross the Hafren ever since Roman times so the initial toll on the new bridge perhaps shouldn't have been that surprising. But in the 1990s the charging system was changed so that, while there was no cost to leave Wales, those wishing to enter had to pay. Within south Wales this just reinforced the longstanding sense of rancour about the toll and its effect on the local economy. The bridge felt less like a gateway and more like a barrier, established and regulated from London – as Wales has been for centuries.

By the early 1990s the old bridge was in need of major repairs and was struggling to handle the increase in traffic generated by the development of the M4 motorway. A new bridge opened in 1996, imaginatively named the Second Severn Crossing. It's the one we are standing under right now. In 2018 it was renamed the Prince of Wales Bridge – a name change that, once more, infuriated many of the people still bristling from the perceived slights of the first bridge.

We leave Black Rock and walk through a small park that leads inland to the village of Sudbrook, located on the cliffs above. Sudbrook was built between 1873 and 1886 with one singular purpose – to house the army of workers required to excavate and construct the Severn railway tunnel. Today, it is an odd mix of new-build houses and old brick row dwellings, but its central feature remains the tall rectangular brown-stone pumping station built by Hawkshaw to control The Great Spring.

Building the 7,000-metre-long tunnel – still the second longest mainland railway tunnel in the UK – was such an incredibly complicated endeavour that it took some fourteen years to complete. It derailed the careers of some of its backers and boosters and it ran ridiculously over budget. Without the ingenuity of Fleuss and the derring-do of Lambert, it might never have been finished.

Even today, the main pumping station at Sudbrook removes some 50,000,000 litres of water from the tunnel every twenty-four hours – some of it being sent to a local brewery a few miles away. Without this round-the-clock operation, the Severn Tunnel would fill with water within twenty-six minutes.

When it finally opened, the Severn Tunnel transformed the transport of goods between England and Wales and ushered in a new generation of economic immigrants at a time when industry in south Wales was booming. It also came to shape the prevailing viewpoint that the best route out of Wales had to be through England – a worldview that would be later cemented with construction of the two Severn bridges. This, despite the fact that the majority of the Welsh coast faces away from England and out to sea, and that for millennia before the people of this land had used sea passages to forge connections with communities all over the world.

We leave Sudbrook and track along the half-moon perimeter of the old Iron Age coastal fort. A newish housing estate sits inland below us. The adjacent field is deluged and I can't help but wonder about the wisdom of building properties on what appears to be an obvious floodplain. It smacks of thumbing a nose at nature and the might of the Hafren. Later, when I get home, I check the map to learn more about this development. One of the main streets running through it is called Great Spring Road. Perhaps the irony was lost on the developers.

Our first walk offered a glimpse of the history and archaeological importance of this fascinating if unfashionable eastern fringe of the Welsh coast. Now I want to dig deeper, to discover what other stories and secrets lie hidden in plain sight. After all, I grew up just down the coast, yet know next to nothing about the history of the Gwent Levels, an expanse of low-lying, mostly agricultural land that occupies the Welsh side of the Hafren Estuary.

Today our planned tramp leads from Caldicot down to Goldcliff nine miles away. The sun is shining and the moody gloom that hung over our walk from Chepstow has lifted.

At the raised, earthen sea wall, the funnel of the estuary widens. Inland, Wentwood Forest and Gray Hill protect the Welsh flank. Over in England, a large storm cloud hovers over the town of Portishead at the mouth of the River Avon. Shafts of rain shoot down out of the cloud – it looks like an alien spaceship about to blast off.

On this side of the estuary, the tide is retreating exposing the jet-black mudflats that it has shaped into orderly terraces. The remaining water slinks its way back out to sea through the gaps. In the distance what appear to be black-tailed godwits take light steps across the mud in search of an invertebrate lunch.

This intertidal stretch of the estuary between Chepstow and Cardiff might not rival Pembrokeshire or parts of Gwynedd for picture postcard beauty, but it is world-renowned for its ecological and archaeological riches, and the insight it has provided into the ways Neolithic, Stone, Bronze and Iron Age societies lived and adapted to a changing landscape.

The southeast part of the Welsh coastline has undergone a mindboggling transformation since (and because of) the last ice age. During peak glaciation (22,000 years ago), sea levels were about 130 metres below their present levels. As the ice started to melt, the sea level rose, but even towards the end of the last ice age some 11,000 years ago the sea was fifty-five metres lower than it is today. What we now know as the island of Britain was still connected to mainland Europe via Doggerland in the east of England. Back then, the Hafren Estuary was a vast low-lying bay stretching as far south as Cornwall in England and Pembrokeshire in Wales. The river flowed through it but only reached the sea at a mountain – what we now call Lundy Island.

The river valley would have been much deeper and steeper than it is today. So, when the glaciers melted and the sea level rose, these valleys quickly flooded and the resulting wider estuary accumulated sediments, sands and peat. These have proved to be invaluable in preserving the archaeological clues of where and how our ancestors once lived.

One man was instrumental in uncovering much of this estuary's archaeological finds. His name was Derek Upton but he was no ivory tower academic. He was a technician at the local Llanwern steelworks and he spent a lifetime walking and exploring the banks and mudflats of the estuary.

Over the years, Upton became an expert on the topography and idiosyncrasies of the Gwent Levels as well as its natural history. He noted the patterns of ancient wooden stakes in the mud – what he perceived to be fishing structures and the outline of huts. From these observations, Upton became convinced he had discovered crucial clues about the lives of our ancestors long hidden in the intertidal sediments.

Some of his early discoveries were dismissed at the time by academics – possibly because they doubted that an 'amateur' who worked at the steelworks could have a better grasp of the past than them. But since the late 1970s, Upton's unconventional but uncanny ability to identify ancient sites has gained respect among a new generation of researchers.

His first major discovery took place in 1979 at Chapel Tump, near Chapel Farm – 100 metres from where we are now. It was here, amid the mud and sediment, that he identified wooden structures later confirmed (using radiocarbon dating) as being Bronze Age. Perhaps these were the same black, blunted stakes we have just passed.

One of his most remarkable finds came in December 1986 when he identified human footprints – both adults and children – in the mud at Uskmouth, a few miles south of Newport. Subsequent research confirmed the footprints were made more than 5,000 years ago.

In 1990 a dig near Caldicot Castle produced a remarkable clue as to how the Bronze Age inhabitants of these shores travelled further afield. There, archaeologists unearthed a large plank fragment from a boat dating back to 1800 BC.

What has evolved, thanks to Upton's meticulous and exhaustive 'hobby', is a new understanding and appreciation of the way humans adapted to living along this lowland coast. Even in our earliest times, people settled close to the sea because of the access and the sources of food it offered. Time and time again, they were forced to retreat and adapt as the glaciers melted and the waters rose.

Over the years, Upton's work both in archaeology and natural history has been celebrated by the local community and global academics alike. Upton, however, struggled through the last decade of his life with a serious foot injury that he had suffered while working at the steelworks. This affected his mobility and his state of mind. On 25 September 2005, Upton left his home with a shotgun and headed out onto the sea wall of the Levels. His body was later discovered on the banks of the estuary that he had dedicated his life to.

Another squall appears on the horizon casting a black shadow over the Mendip Hills south of Bristol. It's still dry and sunny on this side of the estuary, but Andy and I have enough experience of getting drenched to realise it's time to move on. There are still five miles before we reach the village of Goldcliff.

The grass on the raised sea wall has been churned up in the rain so navigating the periodic swing gates on the path is more than a little treacherous – we tiptoe around the slippery sections taking care not to fall into the mudflats below. Fresh obstacles await ahead – a large group of fishermen who have taken up positions all along the sea wall, their rods cast into the murky waters below waiting for a catch.

I'd read about the estuary's rich angling tradition; further upstream at Sudbrook local people have fished for centuries using lave nets – large Y-shaped contraptions that resemble oversized lacrosse sticks. They wade out across the mudflats at low tide knowing that they only have a few hours before the waters rush back in. Their knowledge and mental map of how and where to navigate the mudflats without getting stranded has been passed down through generations of lave fishermen.

Still, it's a bit of a surprise this morning to see such a concerted effort in what, to this untrained fishing eye, looks like a very unlikely hunting ground. I ask one of the fishermen what they are hoping to catch. 'We've all come over from Bristol. We're part of a local fishing club and we heard the cod was good on this side of the river,' he says.

'How is the fishing so far?' I ask.

'Nothing biting yet,' he replies, adding, 'I just hope we're in the right spot.'

Further along the sea wall we meet more fishermen. These are locals and, without giving too much away, intimate that the Bristolians have got their positioning *all* wrong (and aren't about to tell them).

In the distance the village of Goldcliff rises above the flat terrain. Almost 150 years ago, a discovery was made just south of the village that provides an intriguing clue as to how 'the Levels' as we know them came into being.

The exact date isn't known but sometime in November 1878 an inscribed stone was found in the sedimented cliff near the current sea wall at Goldcliff. It read: 'COH I, C STATORI, M… XIMI, P X… XIII S' – Roman Latin script that roughly translates as 'From the first cohort, the century of Statorius Maximus (built) 33½ paces…'

Centurion Statorius Maximus was the leader of the first cohort of the Second Augustan Legion based at Isca – what we now know as Caerleon. It was one of the most important Roman fortresses

Seascape

and one of only three permanent legionary camps in the British Isles. It was home to 5,000 soldiers and was in use for almost 300 years following the defeat of the local Silurians.

Back in the nineteenth century, antiquarians were quick to claim that the 'Goldcliff Stone', as it became known, was a marker of the ancient sea wall built by the Romans. Today archaeologists are less certain. They do know that the stone dates from the late second or early third century and it is most definitely Roman. What they can't say is whether its place of discovery was where the legionnaires started building a sea wall over 100 years before. Sea levels at this time were about one and a half metres lower than today so the Roman wall may well have been built much further out in the estuary.

Wherever its original location, the wall had a specific objective – to hold back the waters to reclaim the nutrient rich marshland for grazing and agricultural production. Once drained, the Romans embarked on a sophisticated construction project, creating a series of ditches, drains and sluices to ensure the salt marshes didn't flood but remained irrigated. Their efforts created a highly fertile new area of the Gwent landscape – providing an agricultural economy for the settlements of Caerwent and Isca. It's the area we now know as the Levels.

The Romans departed UK shores in the early fifth century and the British Isles entered what is commonly, if dismissively, known as the Dark Ages due to the paucity of accurate historical documentation about a period that lasted over 600 years. What we do know is that, with the Romans gone, the sea wall fell into disrepair and the Hafren Estuary once more established its dominance over the Levels. It would remain this way until the Norman conquest of Wales.

The new warrior knights and noblemen who arrived here in

Gwent from northern France, starting in 1067 and reinforced over the next thirty years, sought to consolidate their military strongholds with the soft power of religion. In 1113, Robert de Chandos, the Norman Lord of Caerleon offered the Benedictine abbots of Le Bec-Hellouin, France a tract of coastal land on which to build a priory.

The site of de Chandos' priory was Goldcliff Island, a piece of high ground on the banks of the Hafren Estuary. The land around it would have been *morfa* (salt marsh) along with rough pasture – only suitable for grazing during summer when it was dry. In winter this part of the coast would regularly flood.

De Chandos granted the monks some 200 acres running from Goldcliff to Nash, on the outskirts of what is now Newport. They set about repairing and extending the old Roman sea wall as far south as the mouth of the Usk. To drain and irrigate the marshland, they emulated the Roman engineering and dug a network of straight ditches, about three metres wide, to control water levels. Here on the Levels, they are called reens (perhaps derived from the Old Welsh word for ditch, *rhewyn*). Some of them improved on existing Roman ditches.

The waterways created by the monks followed existing streams and creeks, establishing a pattern of non-symmetrical channels that exist to this day. At least one surviving reen, Percoed on the Wentlooge Levels, is believed to date from Roman times. From here on the raised wall, we can see this irregular patchwork of water channels – think of a wonky game of noughts and crosses – that criss-cross the reclaimed farmland.

Rainwater that flows down from the hills of nearby Wentwood Forest and then collects in the fields of the Levels is removed through shallow surface ditches called grips. That system flows into field ditches and out into the larger reens, which take the water down to the sea wall where it empties into the estuary through tidal gates called gouts. These were designed so that the freshwater flows out at low tide via a flap. Then, when the tide comes in, the sea water pushes against the flap and closes it. Nearly the whole system is powered by gravity and the flow of water alone. In a further feat of engineering ingenuity, the monks controlled water levels within the fields through a set of weirs along the reens known as stanks. These allow water levels to be lowered in the wet winter months and raised in the summer – hence cultivating and preserving the rich grasslands.

The Coast Path leads us towards the headland where Goldcliff Priory once stood. We are navigating a terrain that had first been shaped by humans 2,000 years ago and hasn't changed that much for the last 800. This part of the estuary remains an impressive testament to the achievements of the monks all those centuries ago. Yet, at the same time, it seems a particularly fragile ecosystem. Inland, the grazing fields of local farms are underwater – the heavy rains on recent days clearly asking serious questions of the reens. Water surrounds us on both sides of the path and it doesn't take a great deal of imagination to see it overwhelming these Levels.

The main legacy of Goldcliff Priory is the village of Goldcliff and the vibrant farming community that grew around it to take advantage of the rich, arable land. By the early seventeenth century, the Levels were divided into many parishes and hamlets. This we know because of the tragedy that befell these lowlands on 30 January 1607. On that day, a great flood swept up the Bristol Channel, inundating an estimated 200 square miles of land, killing perhaps as many as 2,000 people on both the Welsh and English sides of the estuary and causing more than £100,000 (approximately £13 million in today's money) of damage.

The Levels were particularly badly affected with more than 500 fatalities. According to one of the pamphlets published at the time, and titled *Lamentable Newes Out Of Monmouthshire*

In Wales, some twenty-six parishes were flooded. Further south down the estuary, the towns of Newport and Cardiff also suffered major damage.

Here in the village of Goldcliff, you can still see the brass plaque that was mounted on the stone wall of the local church in 1609 to commemorate the disaster. The waters reached over seven metres above ordnance datum (AOD) – the standard of measurement when compared to sea level – and overpowered the sea wall.

Climate historians have long sought to understand what could have caused the great flood – the most devastating ever to hit the west coast of the United Kingdom. Could it have been a tsunami? Certainly some contemporary accounts describe giant waves sweeping up the Hafren Estuary. More likely, though, it was a result of exceptionally strong early spring tides. So, with sea levels rising and flood surges increasing, could it happen again and what would be its impact?

Back in 2007, to mark the 400-year anniversary of the Great Flood, a private consultancy created a risk report for today's coastline based on the historical documents of the 1607 storm. It estimated that the likely range of insured losses to the residential, commercial, industrial, and agricultural properties would be between £7 billion and £13 billion. Over eighty per cent of the losses would occur in the inner Bristol Channel, including the cities of Bristol, Cardiff, Newport and Gloucester.

That was nearly twenty years ago when the idea of the Levels being overpowered by the sea probably seemed a little far-fetched. Today, the prospect seems all too real. The most recent scientific mapping and projections suggest that, by 2050, the combination of sea level rise and accompanying seasonal flood surges could breach the current defences if they aren't continually reinforced and improved – inundating the Levels, threatening the livelihoods of the local community and challenging the way life has been lived along this estuary since Roman times.

2

Turning Back the Tide

Walks from West Usk Lighthouse to Cardiff Bay

A weather-beaten Cardiff Council notice tied to a wooden post at Pengam Green roundabout stops Jeff Jones and me in our tracks as we enter the outskirts of Wales's capital city.

It tells us that the section of the Wales Coast Path we are about to walk is closed for emergency flood defence work. It involves shoring up 'the east and west banks of the River Rhymney, to include rock armour revetments, concrete erosion mats, earth bunds [and] a double flood gate.' In layman's terms that means dumping 150,000 tonnes of boulders along this stretch of coast to manage erosion and storm tides.

About an hour before, Jeff and I had set off on what promised to be a pleasant Sunday morning meander from West Usk lighthouse outside of Newport down into the heart of Cardiff Bay. Jeff and I have been going on walking adventures for many years – exploring some of the most beautiful woodlands in Wales and along many majestic beaches. But, deep down, both he and I are particularly attracted to the 'less travelled' walking routes in

Wales; the ones steeped in industrial grit or design quirkiness. We've spent hours analysing maps of Cardiff to figure out the best urban trails so I knew he'd be intrigued to explore this post-industrial part of the south Wales coastline.

We set off in a sharp, powerful gale. Such is the force of the wind that it feels like we are carrying bricks in our backpacks rather than a bottle of water and a slightly stale pasty. This part of the estuary – the Wentlooge Levels – has a real lawless quality to it. Perhaps that just reflects the transient quality of living on the edge of an intertidal boundary.

Just outside the village of Marshfield, about fifty metres out on the mudflats, a hand-painted sign warns: 'WWCA – Private Shooting'. It signals the designated firing area for the interestingly named Wentloog Wildfowling and Conservation Association. It was founded in 1981 to control the growing number of 'cowboy' shooters on the Levels.

'I wonder where the conservation part comes in,' says Jeff as we walk briskly by.

At Lamby Way refuse dump we seek refuge in Parc Tredelerch, once an important medieval farm. The land was recovered from the dump and rejuvenated in the early 2000s to create a park and a mini refuge for a menagerie of mute swans, great crested grebes, herons, coots, mallards and tufted ducks. It feels like a real oasis in this grimy part of Cardiff.

That sense of serenity vanishes as we rejoin the Coast Path at a main road. A steady procession of trucks and vans rush by on their way to or from the rubbish dump. We pass what looks like a shrine to someone who lost their life on this stretch of road – the wilting flowers covered with a sheen of muddy black road backwash.

Now, with the Coast Path blocked, we have no option but to cut back inland into the eastern part of the city. Abandoning the coast feels like a defeat but Jeff has identified a silver lining to raise our spirits.

Seascape

'You know we are just around the corner from the Royal Oak,' he says. 'I used to drink in that pub decades ago and I haven't been there since.'

We head inland through the neighbourhoods of Pengam and Tremorfa that the new sea defences aim to protect. Many of the houses here were built on the site of the old Splott Aerodrome/Cardiff Municipal Airport. Some of the streets are named after parts of the airport and the planes that flew there. There is Runway Road, DeHavilland Road and Hawker Close. Nearby sits Willows High School, named after Ernest Willows, the first person to fly across the Bristol Channel back in 1910 in a home-made airship.

The main bar of the Royal Oak is already full and it's only just turned noon. The core of locals – mainly old men in to watch the early football match – look at us, soaked to the skin and covered in mud from the soggy Coast Path, in bewilderment. This really isn't a pub many hikers visit.

Which is a shame because the Royal Oak is one of the greatest old boozers in all of Cardiff. In the old days, the landlord served Brains beer using a gravity pull system from casks above the bar. Those antiquated methods have long been abandoned but the Victorian wooden bar is still here, as is an homage to one of the pub's favourite sons – the boxer James 'Peerless Jim' Driscoll. His cousin ran the Royal Oak back in the early twentieth century when Driscoll came to fame by winning the Lonsdale Belt – the pinnacle of British amateur boxing. He later bequeathed the belt to the Royal Oak and, in subsequent years, the establishment further cemented its boxing pedigree by developing a sparring gym on the upper floor.

Jeff heads to the bar to get us some drinks and I take a moment to read up about the Afon Rhymni coastal defence work that has brought about this pleasant detour.

A news article says that the project aims to reduce the risk of flooding for 1,116 residential and seventy-two non-residential properties with the hope that the protection will last for 100 years.

I do a quick tally of the potential bill. Even with my rudimentary maths skills, I deduce that, with a published projected budget of £23.5 million, it could cost approximately £200,000 to protect each residential property and the infrastructure that supports them (that's before any budget overrun, which tends to happen with every major public works scheme). The median house price in Tremorfa is £207,000. Roughly speaking, Cardiff's city government has determined it is worth spending the equivalent of the current entire value of the properties to push back the sea over the coming decades.

It is already becoming clear to me that communities all along the Welsh coastline will face similar decisions over the coming years. If it costs more than £20 million to protect just one part of east Cardiff, imagine what the bill might be to fight the sea throughout Wales.

I do some more online surfing about the area we've just walked through. I land upon an article about the old Splott Aerodrome. It says that the first airstrip was constructed back in 1931 on farmland that, in turn, had been reclaimed from salt marsh. You won't be surprised to learn that, soon after opening, the airport flooded as a result of high tides which halted all flights until April 1932.

The more I think about the Afon Rhymni sea defences and the enormous effort to save Tremorfa, a neighbourhood that literally translates as 'town on salt marsh', the more I start to wonder how so much of south Cardiff could have been built in areas that are obvious flood risks.

To answer that question, I'm going to have to take a walk back in time and explore how the coastal part of the city evolved.

So, a few days later, I head back out in search of the Coast Path as it snakes through the streets of Cardiff. I walk through

Llandaff Fields, my local park, towards the Afon Taf (River Taff), cross Blackweir footbridge then continue into Bute Park and the grounds and gardens of Cardiff Castle.

Bute Park is named after the Marquesses of Bute, the most influential family in the development of Cardiff during the nineteenth century and the driving force behind the transformation of the Taff Estuary into one of the world's most important ports. The Bute family made a fortune from the development of the coal industry, partly because they owned thousands of acres of land north of Cardiff where rich coal seams were exploited, and partly because the Second Marquess, John Crichton-Stuart, took an almighty gamble on building a new ocean-going port so the coal could be exported. In the early decades of the nineteenth century, he began the transformation of Cardiff and south Wales by granting mining rights to a new cadre of entrepreneurs. They swiftly, brutally and to great effect turned the once verdant south Wales valleys into a driving force of Victorian industry.

It is lunchtime and the park is full of people enjoying the large expanse of green space that sits on either bank of the Taff. Runners, cyclists and determined walkers jostle for position on the main tarmac path through the park. Others, keen to escape the throng, explore the thin dirt side track that hugs the old, disused, feeder canal. It once carried fresh water down to Cardiff docks to ensure the entrance to the tidal Hafren Estuary wasn't blocked by mud and silt. A small group of teenagers disappear into the bushes and undergrowth by the side of the river – looking for a bit of pebble beach to hang out and smoke weed (judging from the aroma wafting through the park).

Bute Park ends as Cardiff's city centre begins. I head down Westgate Street (so named because it used to be the west gate of the old town) and pause at the corner of Quay Street. Opposite sits the white metal exoskeleton of the Principality Stadium, home to the Welsh national rugby team. Just over 200 years ago, however, this was the main dock for Cardiff's small seaport – and the reason why the name Quay Street endures to this day.

This stretch of the Taff had been navigable for ships since Roman times. The etymology of nearby Womanby Street, which connects Cardiff Castle to Quay Street suggests that Norse traders or Viking raiders once frequented the town. Womanby is said to be derived from the Old Norse word *houndemammeby*, which means 'huntsman's dwelling'. By the Middle Ages, Cardiff's river port was exporting wool and even cannons to England and abroad, but it was still a minor presence compared to the English port of Bristol, situated just a few miles across the Severn channel.

Until the turn of the nineteenth century, the small town of Cardiff was enclosed by a stone wall. The south gate was by St Mary's Church, long since demolished and now the site of the Prince of Wales pub. Below the south gate lay the moors – salt marsh flats that stretched from Afon Rhymni in the east to the Afon Elai (Ely) in the west. The immediate area south was known as the Dumballs and countless lives were lost there over the centuries during high tide floods.

It was the arrival of mass industrialisation that changed the physical layout of Cardiff and reshaped its coastline. The success of the Cyfarthfa Iron Works and its competitors in the industrial valleys town of Merthyr meant that its owners needed fast and efficient ways to transport their products to the coast for export.

A solution was found through the construction of the Glamorganshire Canal – a project championed and funded by Richard Crawshay, the alpha owner of Cyfarthfa. Work began in 1790 and within four years the twenty-five-mile-long waterway had been completed, linking Merthyr to Cardiff. Four years later a new section was added, connecting Cardiff to a new sea lock on the moors south of the town in what is now known as Hamadryad Park. The canal cost over £100,000 and made the old Cardiff quay obsolete. Soon, it would transform the small town of just 4,000 people.

If the iron industry started reshaping Cardiff's coastline, it was coal that radically changed it. The Second Marquess of Bute owned the land south of the old town walls and was keen to exploit it and drive the growth of Cardiff as a port. He was land rich but cash poor so, when he ploughed over £300,000 into the construction of the West Bute Dock, he took on an enormous debt. The dock opened in 1839 but initially struggled to attract enough ships to make it profitable. However, in 1841, a new railway, the Taff Vale line, opened, connecting Merthyr to Cardiff.

The Taff Vale supercharged south Wales's coal exports. Soon Bute's dock couldn't handle the size of the ocean vessels that needed access to the port. So the Second Marquess funded a new, larger forty-five-acre basin known as the Bute East Dock. It opened in stages during the 1850s though Bute didn't live to see his investment – he died in 1848 aged fifty-four. He left a young heir, aged six months at the time of his death. Over the next fifty years, the Third Marquess of Bute would continue his father's drive to establish Cardiff as a world-leading port, building a third dock, known as the Roath, in 1887. In 1896 Cardiff overtook New York as the world's busiest port for exports, dispatching 6.9 million tons to New York's 6.5.

I head up Quay Street and along pedestrianised Queen Street until I reach Churchill Way, a wide boulevard built in the late 1940s on top of the old feeder canal. Today part of the canal is being uncovered as part of the local government's attempt to celebrate Cardiff's maritime heritage.

A new mixed residential and commercial area called the Capital Quarter has sprung up since I last mooched around down here. The name is as bland and boring as its architecture – unimaginative tower blocks masquerading as considered minimalism. It is hard to imagine that this once was Newtown, or Little Ireland as it was known – a raucous, rough and tumble settlement built to house the Irish workers who excavated the new Cardiff docks.

A large influx of Irish immigrants arrived in Cardiff in the wake of the great Potato Famine of 1845. Newtown swiftly expanded to become Ellen Street, North Williams Street, Pendoylan Street, Pendoylan Place, Roland Street and Rosemary Street – just to the south of what is now Cardiff city centre. It was a cramped, overcrowded, unsanitary and challenging place but for the people who lived there it was home and it engendered a strong sense of community. Its most famous son was none other than Peerless Jim Driscoll (of the Royal Oak fame). When he died in 1925, 100,000 people lined the local streets to watch his funeral cortège go by. Today, just an old stone wall remains of the neighbourhood though one of its iconic local pubs, The Vulcan, still survives – albeit as a relocated exhibit in the St Fagans National Museum of History five miles away.

I reconnect with the Coast Path at Brigantine Place, near the original Bute Dock. It follows the feeder canal, taking a series of straight then right-angled routes (think of an aquatic game of Tetris) through an appealing if slightly run-down housing development. The canal water has an algae green tint and the pathway is overgrown in places. Nonetheless, it's an intriguing new route – one I've never fully explored before.

The canal and path emerge into what remains of the once enormous Bute East Dock. Even today, this truncated rectangular expanse of water is impressive. Sunlight glistens off the surface and a cluster of fishermen, each sitting in their own round tent, have cast their rods. I have my doubts about how many fish they'll catch down here but it seems a decent way to spend a day.

As Cardiff docks grew, so did the need for labour. The mid- to late nineteenth century saw an influx of new Cardiffians from all over Wales, England, Ireland, the rest of Europe and the world (notably sailors and traders from Somalia and Yemen). Many settled in Butetown – or Tiger Bay as it was better known – a new urban and commercial development south of the old town wall, right next to the new docks.

Butetown, as the name suggests, was the dream of the Second Marquess of Bute. He harboured ideas of creating an architecturally elegant model community to showcase the success of his new docks. As the port grew in importance, grand houses were built around new squares named Mount Stuart and Loudoun. At first, the community was a mix of merchants, ship-owners, coal brokers and seafarers. But as the port became busier so Tiger Bay became more crowded. The well-to-do decamped to new suburbs like Llandaff and the big houses were divided into multi-family lodging houses to accommodate the influx of new immigrants. The coal ships leaving and arriving in Cardiff docks were often crewed by sailors from the Caribbean, North and West Africa and it was Tiger Bay where many of these men made their homes. By the early twentieth century, more than fifty nationalities were living in this new 'town' including strong contingents from Somaliland, Yemen, Liberia, Sierra Leone and Nigeria.

The extraordinary growth of the docks meant that Bute's new town couldn't accommodate all the people who arrived in Cardiff during the mid- to late nineteenth century. The obvious and quick solution was to build housing on the marshland that surrounded the docks.

This fresh wave of immigrants – many of them from the English West Country and Midlands, Ireland and, increasingly, from across Europe – made their homes in a new, quickly constructed neighbourhood on the banks of the Taff and Ely estuaries. This was land principally owned by Harriet Windsor-Clive, heir of the Earl of Plymouth estate, and the patron of nearby Penarth Docks which had been built to compete with the Butes. Within a few decades it would become one of the most populous parts of Cardiff – known as Grangetown.

It was here, on Warwick Street, where my great-grandfather Axel Frederick Prytz settled in the late 1890s. He was from Landskrona in southern Sweden and came to Cardiff to work as a ship's steward. In 1899 he married my great-grandmother Catherine Ford. Two years later, my grandfather Hubert Frederick Prytz was born.

The new neighbourhood was named after the Cistercian Grange (or farm) that had been located on the West Moors marshland since the thirteenth century. Two separate areas anchored the new neighbourhood. The village of Lower Grangetown was first to grow in the 1860s, centred around Holmesdale Street. Almost immediately it earned an unsavoury reputation. 'It appears from the police reports [that] pigs, donkeys, cows, children and chickens have been in the habit of messing together, in one dwelling, and even in the same room,' reported the *Cardiff Times*, a muckraking newspaper founded in 1865 to challenge the power of the Butes and other major landowners.

The second area to be built on was known as Saltmead. It began as another development for Irish immigrants but much of the land around it was impassable moorland. In 1865, the *Cardiff Times* reported how residents of Canton (a neighbourhood to the north) had petitioned for a new road to be built to help people cross the moor in winter. 'It was very unpleasant now to go to Grangetown across the moors,' said the rector of Canton, Rev'd Vincent Soulez.

In 1875, Grangetown was incorporated into the town of Cardiff, although marshland still separated it from the town centre. The same year, the Afon Elai burst its banks flooding the entire area.

Despite the obvious risks, the development of this precarious landscape continued with more streets being built and families moving in. Some of the building was well planned but much of it was hastily constructed and ramshackle. Many of the dwellings were dangerously overcrowded with multiple families living under one roof.

More flooding in 1883 prompted the construction of a new system of clay banks but these sea defences were precarious at best. In the early 1890s, local councillor Samuel Arthur Brain (founder of Brains Brewery) lodged an official complaint about the 'swamp' in Saltmead. Over time, his concerns became known as the Saltmead swamp scandal and prompted an official investigation. According to the *Western Mail* newspaper, Cardiff's medical officer found drainage and health problems due to some homes being built on clay, with stagnant water under the floors. It questioned the integrity of both the planning process and building work, writing that the 'poor were being choked out of existence in the Saltmead swamp.'

'The fool who built his house upon the sand was a wise man by comparison with the builders of certain streets in Saltmead,' the paper concluded.

The fall of Cardiff docks was almost as rapid as its rise. At the start of the twentieth century, Cardiff was considered one of the most important ports in the world. At the nearby Coal Exchange, in the heart of Mount Stuart Square, brokers gathered every morning to set the price of shipping coal for the day. In 1904, the first £1 million coal trade was recorded. Such was the wealth being created each day that the brokers were known to play skittles with the empty champagne bottles they'd drunk.

The docks reached their peak in 1913. That year, nearly eleven million tonnes of coal were exported from Cardiff. The only global port that shipped more was nearby Barry Docks, built in 1880 to compete with Cardiff. But then came the Great War, shattering the coal export market on which south Wales mines depended. By the 1920s, cheaper sources of coal from German and US mines made Welsh coal uncompetitive. The entire global industry also was undercut by the growing importance of oil as a heating and transportation fuel. By 1932, coal exports from Cardiff had fallen to below five million tonnes. Cardiff docks experienced a brief reprieve during the Second World War but its central reason for existing – to export coal – had disappeared. The Bute West Dock was closed in the years after the war ended. By 1970 the Bute East Dock also had been decommissioned.

The docks that I remember as a child in the 1970s were very much an industrial wasteland – but an accessible one. The East Dock was used periodically to host speedboat racing that my family would come to watch. Other times, we would drive down here on a Sunday afternoon to watch ships unload their cargo in what must have been either the Roath or Alexandra basins – the last two functioning docks. I remember being surprised that we could drive our car almost next to these big ships. Most of all though, I remember how grey, intimidating and yet intriguing this down-at-heel industrial hub felt back then. And how the small boats moored in the harbour would be stranded at angles on the mudflats when the tide retreated – a reminder of how, throughout Cardiff docks' highs and lows, the sea remained in control.

That balance of power would change in the 1990s with the development of Cardiff Bay – as bold a capital investment project as anything those nineteenth-century coal barons ever attempted.

The first move to revitalise the former docks came in the late 1980s when South Glamorgan County Council, the regional authority at the time, decided to relocate its headquarters from the city centre down to the banks of the Bute East Dock. The building it moved into was a new commission – a modern pagoda design that was a striking departure from the grand, classical nineteenth-century stone architecture of the Bute era. It was designed by the county architect John Bethell – my friend Andy's dad.

The council's decision to relocate to the depressed docklands was a shock to many (as was the new-look headquarters) but it soon sparked new business and government interest in redeveloping the area. This was led by Nicholas Edwards, the Welsh Secretary in the Conservative UK government at the time. Edwards was of Welsh heritage but was very much of the English establishment. As a Conservative in the early 1980s, he held a great deal of power over a nation that was overwhelmingly Labour voting. According to a local business figure at the time, he acted like a 'Roman proconsul'. Edwards was, in short, as much of a patrician figure in Cardiff's development as the Marquesses of Bute had been before. Arguably he would have as big an influence.

In 1987, the Cardiff Bay Development Corporation was established to transform the old docks into a completely new geographical and cultural space. It promised to bring new businesses to south Cardiff, to bring tourism to this part of the coast and revitalise Butetown through new job opportunities that 'would reflect the hopes and aspirations of the communities of the area.'

What Edwards cared most about, though, was the construction of an opera house, one that would realise his dream of making Cardiff a citadel of highbrow culture. 'I thought that a centre for the performing arts should form a central component to the Cardiff Bay plan,' he wrote in his autobiography. The result would be the Wales Millennium Centre – the truly impressive spaceship-like building that sits today at the top of the old Bute West Dock.

I had walked by its slate-fronted façade just minutes before. Two lines by Welsh poet Gwyneth Lewis are inscribed in its bronze-coloured dome. They read:

> CREU GWIR
> FEL GWYDR
> O FFWRNAIS AWEN
> IN THESE STONES
> HORIZONS
> SING

The Millennium Centre is one of my favourite buildings in all of Cardiff. It dominates the skyline for miles around – especially when lit up at night. Today it's the home of the Welsh National

Opera but it's not the opera house that Edwards campaigned for. He had supported a design put forward by the renowned architect Zaha Hadid, which failed to get funding due to what some in government considered an overly radical design. Hadid blamed the rejection on institutional racism and sexism.

Ultimately a Welsh architect, Jonathan Adams, was commissioned to design the new cultural centre. He took inspiration from many parts of the landscape but was particularly inspired by the coastal cliff faces of Ogmore and Southerndown, a few miles down the Glamorganshire coast.

Edwards's personal vision for the opera house may have been dashed but his greater plan for Cardiff Bay – a half-mile-long barrage between Queen Alexandra Dock in Cardiff and the entrance to Penarth Docks – will prove his lasting legacy, even if its merits continue to divide people in the city to this day.

The idea for the barrage was initially formed from an argument over whether the tidal mudflats of the Taff Estuary were the right 'look' for the new Cardiff Bay. One of Edwards's advisors suggested the creation of a permanent freshwater lake to attract investors. The Welsh Secretary keenly embraced the idea.

The barrage faced criticism from many sides. The British Prime Minister, Margaret Thatcher, opposed it on budgetary grounds. The local Labour Party MP, Rhodri Morgan, also railed against the high cost of building the barrage but he was particularly vehement about the potential environmental damage posed to wildlife and bird feeding grounds on the mudflats. Local residents of Grangetown were fearful that the creation of an inland lake would increase the risk of flooding to their homes. Those living in the old Tiger Bay, meanwhile, feared their community would be neglected in the rush to appeal to the financial community and commercial property developers.

Ultimately, after much fighting and a great deal of lobbying in the UK parliament, the plans for the barrage were passed. It was built over a period of six years with its grand opening taking place in 2001. The result of hemming in both the Taff and the Ely rivers is a mile-square body of freshwater whose water levels are regulated by a series of sluice gates. Boats can enter and exit the bay using a series of locks built into the barrage while a 'fish pass' allows breeding salmon to swim upstream. To mitigate the environmental impacts of reshaping the coast, the Uskmouth Nature Reserve was created just east of Newport – wildlife offsetting for want of a better term.

In terms of pure economics, Cardiff Bay has been a qualified success. It created a restaurant scene and attracted many more tourists to the city. It became the home not just for the Millennium Centre and BBC television studios but also the Senedd – home of Wales's devolved government. But, just as many local people suspected, the new wealth failed to reach the Butetown community that had contributed so much to the growth of Cardiff. The new physical layout and the economic development of the bay balkanised and defanged Tiger Bay – isolating its community and robbing much of what had given it character.

The sun is gracing Cardiff Bay as I walk over the barrage towards the town of Penarth. At the pirate ship playground halfway across the barrage, I take a moment to look back across the bay towards the city. The metal roof of the Millennium Centre shines in the sun, the red brick Pierhead building that once greeted sailors entering the West and East Docks looks proud and majestic while the white St David's hotel, designed to look like a giant sail, promotes a modern if weather-beaten elegance. Small sail boats navigate the mini whitecaps in the bay. The only lasting signs that this once was a working dock are the five decaying wooden 'Dolphin' mooring platforms that jut out from the water near the old black-and-white Norwegian church, built in 1868 to provide spiritual support for sailors while visiting Cardiff.

Despite its faults and inequalities I have to concede that, on days like this, Cardiff Bay and the barrage have a definite appeal. And despite the very valid concerns of Grangetown residents that the bay would raise groundwater levels and threaten their properties, it has so far helped prevent the regular seasonal flooding that used to afflict the neighbourhoods built on the Ely and Taff floodplain. How it copes with the next generation of sea and river level surges – both from the Hafren and the Afon Taf and Elai, which carry water down to the bay from the surrounding hills – will determine the next chapter in Cardiff's complicated relationship with its coastline.

3

Castles in the Sand

Walks from Penarth to Barry, Cold Knap to Ogmore and Newton to Kenfig

In early May 1897 an enthusiastic young Italian man named Guglielmo Marconi travelled to Lavernock Point, a mile or so west of Penarth.

Marconi had been born to a wealthy family in Bologna, Italy in 1874. He showed a fascination in science from an early age – specifically conducting experiments in electro-magnetic waves. In 1895 Marconi had sent a Morse Code message over two miles across the family's country estate. Delighted by this achievement, he approached the Italian government but they were unimpressed.

Undeterred, Marconi headed to London. He realised that his work on messaging systems could be of great importance to shipping and Great Britain ruled the waves at this time.

Marconi was introduced to Sir William Preece, Chief Engineer to the Post Office and an inventor in his own right. He immediately saw the value of the young Italian's work and assigned him an assistant named George Kemp, a post office employee from Canton, Cardiff. It was Kemp who suggested Marconi test his invention by transmitting from Lavernock to Ynys Echni (Flat Holm), an island roughly four miles out in the mouth of the Hafren.

Until Marconi turned up, Ynys Echni was best known as a place of hermitage. In the sixth century a prominent local Celtic saint, Cadoc, travelled to the island to meditate and escape the pressures of the Dark Ages. The Anglo-Saxon word for the island was Bradanreolice – derived from an old Irish word for graveyard or burial ground. A few centuries later, Viking sailors are said to have used Echni as a base to launch raiding parties on the mainland. (They gave it its English name as holm is a Norse word for a small island.)

On arrival on the south Wales coast, Kemp, accompanied by his nephew, Herbert, sailed out to Ynys Echni and erected a 100-foot radio mast. Marconi, situated at Lavernock, did the same. On 13 May 1897, after a number of failed attempts, Marconi typed a simple Morse Code message into his transmitter. It read: 'Can you hear me?'

Minutes later the printer attached to his equipment returned a message from Kemp. It read: 'Yes, loud and clear.' The first radio telegraph message across water had been successfully sent.

Marconi continued to experiment and build on this achievement until, on 12 December 1901, he succeeded in sending the first radio transmission 2,000 miles across the Atlantic Ocean. It disproved prevailing scientific consensus that radio transmissions couldn't carry more than 200 miles due to the curvature of the planet. In an age where humans had yet to fly, and where all transatlantic communication had to be delivered through underground cables, Marconi had shown how to break the shackles of the sea.

His legacy reshaped the world of radio communications, ushering in the electronic age that drove so many of the twentieth-century inventions we now take for granted. So you'd think he might deserve a more impressive acknowledgement of his achievements than a single plaque mounted in the stone wall of the local churchyard near Marconi Holiday Village, a static caravan park that now occupies the Lavernock headland.

Perhaps, though, this was as much as he deserved. Later in his career, Marconi returned to Italy where he joined Benito Mussolini's Italian Fascist Party. Il Duce made Marconi a member of the Grand Council of Fascism and put him in charge of the Academy of Italy. In 2002 historical documents were unearthed showing how he blocked all Jewish candidates from joining the Academy at Mussolini's behest.

The sea is calm today – its light muddy brown colour resembles a very large crème brûlée. Jeff has rejoined me and we are looking out over Lavernock Point just as Marconi would have.

It feels good to be out on this bright early spring day. I'd been stewing in my own head for the past few weeks – letting the little everyday problems that normally I'd dispatch without too much thought build up and calcify in my brain. I knew that getting out into nature and walking would be the solution to my fogginess. Walking calms my mind and clears my head. It gives me space to think and it feeds my creativity. But I'd fallen out of the habit of walking every day during this particularly miserable wet winter weather. I had to drag myself out on the first walks along the estuary. Now I was out here, where the Hafren starts to open out into Bridgwater Bay, I was glad I hadn't given in to those demons.

The Coast Path wants us to divert inland, hugging the perimeter of the holiday village. Except that I've found a short cut, a public footpath along the cliffs that leads in front of the static caravans overlooking Lavernock Bay.

Jeff, after many walks with me, is wary of any deviation I suggest but he can see the logic and appeal of bypassing the caravan park.

Even if it means ignoring the hand-painted wooden sign directly in front of us which reads 'Dead End'.

'Probably one of the local caravan owners trying to protect their exclusive view of the bay,' says Jeff. I concur and we stride on with a right to roam sense of righteousness for about 100 metres until the path disappears off the side of the cliff. Erosion has robbed us of the short cut. We laugh at ourselves as we backtrack to the official detour – even more so when we pass the 'Dead End' sign. On this side is written: 'Told you'.

We push on past the village of Sully until the path diverts once more from the coast (already a slightly annoying trait of a national coast path). Instead, it leads inland on first a cross-country and then urban odyssey in order to circumvent Barry Docks which is off limits to pedestrians. It seems a shame, though, on such a nice afternoon, to abandon the sea so Jeff and I opt to follow a less trafficked trail that hugs the rocks. It's just the sort of off-the-cuff walking decision that sometimes pays off. We follow the trail through a shallow wooded glade on the edge of a stone beach before emerging onto a rugged escarpment of rocks. In front of us is a clear, crisp view of the water.

This is completely new terrain to me. It feels lost in time. With good reason, we soon discover.

Confused by where we've ended up, I consult my map. We are walking in an area known as the Bendricks – a section of coastline that neither of us have heard of. We sit on the cliff edge, with the waves not exactly caressing the rocks but hardly smashing them either, and search online for more info.

'Bloody hell, this is dinosaur country,' says Jeff, looking up from his phone.

The Bendricks is a stretch of prehistoric carboniferous limestone rocks dating back to the Triassic period (immediately preceding the Jurassic). They are a mix of red and buff-coloured siltstones and mudstones with greyer coloured sandstones. Geologists have determined this was once a desert floor close to the edge of a large lake or sea. The land was inhabited by a range of reptiles, including some of the earliest dinosaurs.

It was here in 1974 that palaeontologists first identified three-toed dinosaur footprints dating back 220 million years. They were made by theropods known as Grallators. More footprints were discovered over the years but then, in 2021, four-year-old Lily Wilder stumbled upon a perfectly preserved Coelophysis dinosaur footprint while out for a walk with her father. The National Museum of Wales described it as 'the best specimen ever found on this beach'. We are stunned to learn this and spend the next twenty minutes picking our way over the rocks looking for our own fossil find only to come up empty handed. No big surprise, we both agree. 'It's tough to outdo a small child when it comes to dinosaur knowledge,' says Jeff.

The next day, Jeff and I head to Barry Island on the other side of the docks. We skip the muted delights of the out of season pleasure park and start instead at Cold Knap where the Coast Path will take us to the town of Llanilltud Fawr (Llantwit Major) just over ten miles away.

Barry Island, as most people steeped in 2000s popular culture know, is the fictional home of *Gavin and Stacey* and the sitcom has turned what was a run-down beach resort into an ironic icon attracting tourists from around the world. Today, locals jokingly call it 'Barrybados'.

Barry owes its name to another Celtic saint named Baruc. Like most tales surrounding the Welsh saints of this period, it involves a supernatural element that reflects just how much ancient Celtic Druidic beliefs about the power of nature were interwoven into the stories of the new Christian church.

This one involves Cadoc, the sometimes hermit abbot, who we met earlier. He had been on one of his periodic prayer and contemplation retreats on Ynys Echni and had taken two followers with him – Baruc and Gwalches. Baruc was tasked with carrying a prayer book that was of great value to Cadoc. But when they returned to the mainland, Baruc realised he had left the book on the island. Cadoc sent him back to retrieve it.

On the return journey from Ynys Echni, Baruc's boat was caught in a violent storm and he drowned. His body washed up at what is now Jackson's Bay – a mile east of where Jeff and I are walking. Cadoc's prayer book, however, was lost to the sea and he was distraught (let's assume he showed the same concern for Baruc).

That night, having buried Baruc in a grave above the beach, the holy men fished for dinner and caught a very large salmon. They cooked it and Cadoc was responsible for serving the dish. As he cut into the fish, he made a miraculous discovery – the salmon had swallowed his prayer book and it was intact. To celebrate the miracle, the holy men built a chapel in honour of Baruc and for many hundreds of years afterwards it was considered an important place of pilgrimage for Christians – so much so that four trips to Baruc's grave were considered the equivalent of one trip to Rome (and a good deal cheaper). The chapel survived until the seventeenth century when it was buried by sand. In the late nineteenth century, antiquarians excavated the site and found more than 400 skeletons – perhaps devotees of Baruc who had requested to be buried at this holy site.

In the early centuries of Celtic Christianity, islands all around the coast of Wales were seen as important and sacred places. Ynys Enlli (Bardsey) off the Llŷn Peninsula coast was known as the Isle of 20,000 Saints. Caldey, an island off the coast of Tenby in Pembrokeshire, is where Pŷr established an abbey in the sixth century. Ramsey Island, close to St Davids Cathedral on the far west coast of Wales, is where Justinian became a hermit only to be beheaded by some fellow monks who had tired of his draconian approach to life. (Though, even in death,

Justinian still had the wherewithal to pick up his own head and carry it across the water to the mainland so he could be buried in peace and in one piece.)

Our climb away from the pebbled beach at Cold Knap is steep and treacherous – mainly because we bought takeaway coffees at a local café and are trying not to spill them as we walk on wet grass. The weather is blustery and the sea has turned from crème brûlée to a colour more resembling sticky toffee pudding. The waves smash against the pebbles of Porthkerry beach dragging the smaller ones back into the sea before propelling them forward once more – as if the sea is toying with its prey.

We make good time – nothing like rain running down your neck to make you pick up the pace – and soon reach Rhoose Point, the southernmost location in Wales. A four-metre-tall menhir celebrates this geographical landmark – it's like something Asterix and Obelix might have left behind if they'd been walking the path. Behind us, next to an old disused and flooded quarry, lies a housing development, part of the bedroom community expansion of Rhoose that has taken off in recent years. Jeff and I look at each other and say in unison: 'flood risk'.

At Limpert Bay, west of Aberthaw, we stop for a break. The sky has cleared and the sun offers just a hint of warmth – though the stiff, cool sea breeze makes sure it doesn't get above its station. It is only March, after all.

A mile or so inland lies the village of St Athan – named after an Irish holy man, Tathan, who founded a religious settlement there in the late fifth or early sixth century. Tathan is considered Cadoc's mentor. The son of an Irish king, he relinquished his title after a visitation from an angel (or so the story goes). He travelled to Wales with eight disciples and sailed up the Hafren Estuary to the old Roman stronghold of Caerwent. There, he established a seat of religious learning and pious wisdom that won the respect of the Celtic hierarchy including King Caradoc (supposedly one of Arthur's Knights of the Round Table) as well as King Gwynllyw (the Newport ruler whose outlaw exploits in the Bristol Channel saw him adopted later as the patron saint of pirates). Gwynllyw sent his son Cadoc to study under Tathan and it was in Caerwent where Cadoc's holy powers were honed.

We sit in a sheltered spot by the rocks and soak up our surroundings. There's a sense of peace but also potency to this part of the coast. The jagged cliffs provide drama while, just off the beach, the green fields are inviting, comforting almost. If only we didn't know just how waterlogged they are.

Perhaps it was this combination of the sublime and the bucolic that attracted early Celtic saints like Tathan and Cadoc to settle on this part of the Glamorgan coast. Both helped establish its spiritual pedigree – the latter's religious base was at Llancarfan, just three miles away. But it was another holy man, Illtud, who would make south Wales an international destination for spirituality and Celtic culture after he settled in Llanilltud Fawr, a mile or so down the coast. Today Illtud's abbey is generally acknowledged to be the first centre of learning ever established in the United Kingdom – predating both Oxford and Cambridge universities by more than 500 years.

Llanilltud Fawr is notable not just because of the sheer number of learned people it attracted – more than 1,000 scholars studied there at its height during the sixth century – but because of where they came from. Those connections tell us much about the way that life and thinking in Wales influenced, and was influenced by, the larger Celtic world of Ireland, Cornwall, Brittany, Northern Spain, and beyond.

Before the Roman invasion and occupation of Gaul, Spain and the British Isles, Celtic people had travelled and traded along the Atlantic coast and into the Mediterranean for thousands of years. They navigated a network of maritime pathways known as the Western Sea Routes. These had shaped migration patterns and enabled the exchange and spread of new ideas, technology and language. Raw materials such as copper ore from Y Gogarth

(Great Orme) in north Wales and Mynydd Parys in Ynys Môn were shipped south while jewellery, pottery and other finery came north.

Roman dominance of land and sea put an end to this trade but, as their influence began to fracture during the third and fourth centuries, these ancient sea routes became popular once again. By the fifth century, holy adventurers from Ireland, known now as *Peregrini*, were helping spread the teachings of the early Christian church all along the Atlantic coast as far south as Galicia – notably the radical new belief in monasticism that spread from Brittany in western France.

Illtud, like many of the Celtic saints, stumbled into a holy life. As with Caradoc, he is said to have been one of King Arthur's knights. After nearly dying during a hunting expedition, Illtud was convinced by Cadoc that God had saved him. Indeed, the story of Illtud seems to be part of a recurring theme in medieval Arthurian mythology where knights turn their back on war to embrace the ascetic life of a monk.

Illtud's monastery quickly grew in numbers and importance. Would-be and some already storied holy men travelled to study at Llanilltud Fawr and then exported their ideas back across the Atlantic coastal nations. Dewi Sant remained in Wales (and would become its patron saint) while others, like Samson, would end his days in Dol – a part of Brittany where another famous Welsh saint, Teilo, also spent time.

Sitting on the beach, I try to imagine what this coastline must have been like back when some of the most influential characters of early Celtic Christianity visited these shores. How what is now just a tranquil, rural part of the south Wales coastline would have been a hotbed of intellectual discussion. How the name Llanilltud Fawr would have been known, respected and revered throughout the post-Roman Celtic world. And how the historical pedigree of this place makes a mockery of the idea that, as modern Wales was starting to take shape during the post-Roman age, it was solely defined by its relationship to the Anglo-Saxon lands to the east.

The next day Jeff and I pick up the Coast Path at Llanilltud Fawr beach. Andy has returned to the fold and brought his friend Al along for the trek. Al is a local lad – he grew up here and spent much of his childhood exploring this part of the coast. He is keen to get reacquainted with some of the trails he walked as a teenager.

Our route takes us along The Glamorgan Heritage Coast – so named back in the 1970s because of its geological, ecological, archaeological and historical significance and to attract tourists into a part of Wales that wasn't on the normal beach/mountain/castle itinerary.

This whole area of coastline was the haunt of pirates back in the late Middle Ages. We reach Tresillian Bay, where one notorious fifteenth-century outlaw, known as Colyn Dolphyn, met his fate. Dolphyn was a Breton pirate who operated from a base on the island of Lundy, once the Ice Age mountain but now just a slither of granite situated about thirteen miles off the English coast in the Bristol Channel. He had already earned himself a reputation for pillage and plunder but his most audacious attack took place in 1449 when his ship captured Sir Harry Stradling, the son of Edward, a powerful local member of the aristocracy and High Chamberlain of South Wales.

Realising the value of his prize, Dolphyn held Sir Harry prisoner aboard his ship, the *Sea Swallow*, for nearly two years before the Stradling family were able to pay the ransom of 1,000 marks by selling two manor houses.

Dolphyn continued to terrorise sailors up and down the Bristol Channel until, one day, his ship was caught in a storm and ran aground here at Tresillian Bay – just half a mile from St Donat's

Castle, the Stradling family manor (and the one they hadn't had to sell). Alerted to the shipwreck, Edward Stradling gathered a posse of men and found Dolphyn hiding in a cave at the edge of the bay. According to one telling of his demise, Stradling buried Dolphyn up to his neck in sand and watched him drown as the tide flowed in.

The goings-on here were hardly isolated incidents of smuggling and piracy along this coast. During the sixteenth century, as the Marcher lords started to lose their grip on a part of Wales that they'd controlled since the Norman invasion, this southern coastline developed a reputation for lawlessness, subterfuge and as a base for pirate raids that reached as far afield as Spain and Portugal.

One reason for this was that, even though Marcher power was fading, Wales didn't officially become part of the English state until the Acts of Union were completed in 1542. Even then it took years to integrate the new laws, not least because local lawmakers and officials were slow to accept the changes – many of which undermined the considerable power they'd once had. A formal customs system for regulating, policing and taxing goods being shipped along the Welsh coast didn't begin until the 1560s, when the rights to impose and collect duty on foreign trade were removed from the Marcher Lords who had previously held them.

Local reticence over sending duties to the English crown meant that the new laws were haphazardly applied at best and actively ignored or undermined in many instances. Just the sort of environment that was perfect for a particular type of sailor willing to take the risk of dodging the law or flagrantly breaking it.

No man more embodied this era of Welsh piracy than John Callis (aka Callice) who hailed from Tintern on the bank of the Afon Gwy, just above Chepstow. He operated from a number of hideaways along the south Wales coast and, during one single year, brought a captured Spanish ship into Cardiff, a Breton ship into Penarth and another into Newport.

His reputation was so fearsome that, in one instance, the French ambassador complained to Francis Walsingham, principal secretary to Queen Elizabeth I, that Callis 'tortured the men and mariners with extraordinary cruelties'.

Despite these failings, Callis had powerful allies in south Wales who protected him because he served a purpose. Indeed, though pirates were active on both sides of the Bristol Channel during this time, it was their close relationship to local Welsh authorities that attracted the ire of the English crown. None more so than Callis.

The pirate's central benefactor seems to have been the powerful Herbert family – one of whom, Nicholas, was the Sheriff of Glamorgan. This merchant family had good reason to resent the new English customs systems as they stood to lose a great deal of money based on the goods they shipped abroad and imported. Some of their land had earlier been seized by the crown and customs documents from the period suggest that the Herberts regularly ignored paying royal customs duties. (Academics believe that the failure to declare goods to customs authorities was endemic among Welsh merchants – suggesting that the actual volume of trade from Welsh ports during this time has been vastly underestimated.)

John Callis didn't escape the law forever. He was captured in 1577 and, in that fine pirate spirit of solidarity, tried to buy his freedom by turning on his fellow outlaws. One account says he was released on 14 July 1578 thanks to his friends in high places (though another says he was hanged in Wapping). That same year, another Edward Stradling, a direct descendant of Colyn Dolphyn's hostage Harry, was put in charge of cracking down on piracy along the south Wales coast.

Seascape

We soldier on, walking along the concrete seafront of the Stradlings' ancestral home – St Donat's Castle. For a short period starting in 1925, this grand home, dating back to the twelfth century, was owned by the US newspaper tycoon, William Randolph Hearst. Quite why he was so keen on this part of south Wales isn't clear. Some think that, having built his Spanish-style estate at San Simeon in California, Hearst simply wanted to own an ancient British castle. Whatever the reason, the tycoon purchased St Donat's sight unseen for about $130,000. He spent lavishly on refurbishing the property and had many friends visit. The castle was said to have hosted some of the most flamboyant and outrageous society parties ever seen in these parts. Hearst himself, though, only spent a total of four months there.

Memories of the Stradlings and Hearst have long faded but St Donat's still has an aristocratic air. Today it is the home of Atlantic College, an elite higher education facility that attracts young highflyers from all over the world, including a fair few European royals. Here they hone their seafaring skills when not studying.

'Those pirates could make a fortune off this lot,' says Jeff as we pass the sea gate.

We make good time and soon reach the whitewashed Nash Point lighthouse.

'I used to listen to the foghorn sounding from the lighthouse when I lay in bed as a kid,' Al recalls. Back then it was still operated manually – the last in Wales to be de-manned back in 1998.

Al was a talented athlete – an international hurdler – in his youth and even today, at sixty, he has a Tigger-like bounce to him as he walks briskly along the mossy grass that layers the cliffs.

As the rest of us keep up a steady if sedate pace, carefully traversing the steep dips and climbs of the path, Al darts his way down the hill, springing from tussock to tussock seemingly without a care for his lower limbs.

'It all looks very good until he breaks his ankle,' says Andy, without the slightest hint of jealousy.

Castles in the Sand

I remember bringing my children, Dylan and Zelda, to Nash Point when they were little kids. Zelda used to hate walking over the loose pebbles to get down to the beach so I would carry her on my back – hoping the stones didn't give way under my feet as we walked. At the end of the day, my wife Jowa and I used to encourage the kids to run up the steep slope and back down again in the hope we could exhaust them so they would fall straight to sleep when we got home. It worked every time.

The beach has always been a favourite of mine. I love how the retreating tide reveals multi-layered flat rocks that stretch out to the next bay and beyond. But it can also be a dangerous place because of regular cliff rockslides – a result of the coastal erosion that eats away at this geologically important part of the coast.

The cliffs here were created more than 300 million years ago when all of this area would have been sub-tropical ocean. They consist of layers of hard grey carboniferous limestone mixed with softer shale and conglomerate that is coloured white, blue and gold. Just like the Bendricks for dinosaurs, this stretch of coast is a treasure trove for aquatic fossil hunters. Trapped in the sand and mud that have long turned to rock are crinoids, brachiopods and corals. But the multi-layered strata are particularly vulnerable to the strong Atlantic storms and the high tidal range both of which erode the softer shale, undercutting and exposing the belly of the harder limestone.

Which is exactly what has happened to the cliff directly next to the stream running down into the beach. It looks like a giant has taken a large bite out of it. Layers of rock halfway up the cliff have been chomped away at – leaving the top sections hanging over the stone beach with no support. It's only a matter of time before the sea winds further erode these rocks and the entire top part of the cliff collapses.

Sometime in the early twelfth century, the new Norman rulers of Wales constructed a castle in the strategically important location of Kenfig, a couple of miles west of modern day Porthcawl. Its purpose was to protect the nearby Church of St James community and to repel Welsh attacks that were launched from the uplands of Margam Mountain a mile away.

By the middle of the fourteenth century, Kenfig was a substantial borough of perhaps 800 people. However, just a century later, the town was all but abandoned – having come under constant attack not from the Welsh but by the sea in the form of huge amounts of sand that accumulated in the town and forced people to give up their homes. The following year the remaining inhabitants were instructed to move to nearby Pyle, where a new settlement was developing. By the 1530s the travel writer and map maker John Leland described 'a village on the east side of Kenfik, and a Castel, both in ruins and almost shokid and devoured with the sandes that the Severne se there castith up'.

Merthyr Mawr, further east down the coast, has a similar tale. There, a village named Treganlaw was said to have been swallowed by the sands around the same time that Kenfig disappeared. Some nights, apparently, you can hear the ghosts of the old village crying out for their lost homes.

My plan today is to find the ruins of Kenfig Castle. My OS map tells me they lie less than a mile inland from the Coast Path amid the sand dunes of Kenfig Burrows. Andy, Jeff and Al are with me as we leave Newton Beach and head west past Rest Bay and Royal Porthcawl Golf Club.

The Coast Path hugs the side of the golf course until we reach a small headland called Sker Rocks. The wind howls off the sea here and we stop to grudgingly appreciate how wild this part of the coast can be when the weather closes in.

And deadly. In the distance, about two miles off Ogmore beach, lies Tusker Rock – a 500-metre-long outcrop that becomes submerged just below the water line at high tide. Over the centuries, it has claimed the lives of many sailors whose boats have wrecked there.

Here, at Sker Rocks, we are at the site of one of this coast's worst shipwrecks – the SS *Samtampa*. On 23 April 1947, the *Samtampa*, a Liberty ship built by the US Navy in the Second World War, developed an engine fault as it was heading east towards Newport. Its captain decided to anchor in Swansea Bay to carry out repairs but, in the early hours of the morning, the starboard anchor failed and gale force winds propelled the *Samtampa* onto the rocks at Sker Point.

A lifeboat was launched to help but it too was overwhelmed by the storm and capsized. By evening, the *Samtampa* had split into two. Within hours, the ship was destroyed. All thirty-nine crewmen of the *Samtampa* died that night as did the eight lifeboatmen. Local residents parked cars on the shore and turned the headlamps on to locate the bodies washed up on the rocks.

I know the ruins of Kenfig Castle are close by. We just have to find them. The long stretch of Kenfig Sands spreads out to our left while, to our right, are the Burrows – the maze-like set of sand dunes and pools that include Kenfig Pool, the second largest freshwater lake in south Wales.

These burrows are part of the greater Kenfig National Nature Reserve that covers 1,300 acres between Porthcawl and Port Talbot. They play a key role both in shielding the coast from storm surges and supporting many diverse forms of flora and fauna including the rare fen orchid.

The protection they offer today might be considered ironic given that they were mainly formed during a 200-year period between the thirteenth and fifteenth centuries – the result of wild storms and tidal surges that created westerly wind-driven littoral sand dunes at Merthyr Mawr, here and at other locations up the Welsh coast. Before the sands accumulated, both Kenfig and Merthyr Mawr were shallow river estuaries formed by glacial movements from the Bannau Brycheiniog (Brecon Beacons) and Carmarthenshire Fans miles to the north.

I consult the map once again. We need to head inland towards Kenfig Pool, navigate our way around its southwest perimeter and then go further into the dunes where the ruins lie.

I lead the way into the dunes and follow a series of mini pools. The sandy ground gets soggier and marshier the closer we get to Kenfig Pool. Suddenly we are at its edge – only a gap in the trees and the tall yellow reeds surrounding the pool give away its location. The water is so well disguised that we could have walked for hours and not found it.

A low mist hovers above the silver-grey water so that we can only just make out the far side. Scientists believe the freshwater lake – a haven for migratory birds and home to a large population of caenid mayflies, bristleworms and pea mussels – was created when the sand dunes swallowed the town. They think the Afon Cynffig once flowed directly south through here to its mouth at Sker Rocks but that the sand dammed its course. The lake, at its deepest, is only about eight metres.

Local legend is more colourful. One tale speaks of a deadly whirlpool that drowns anyone foolish enough to venture into the middle of the pool. Another story – believed by many in the centuries immediately after Kenfig disappeared into the sand – is that a great city lies at the bottom of the lake. It was swallowed by the sea as an act of vengeance after a young suitor committed murder to steal the riches he needed to woo a wealthy man's daughter.

Kenfig and Treganlaw aren't the only reported sunken cities in these parts. Medieval stories speak of a bridle path running from the Penrice Estate on the Gower Peninsula across Swansea Bay to Margam Abbey – just a mile or so from here. And fishermen have long claimed that the bay holds the remains of Grove Island, or Green Grounds as it was also called, which was swamped and washed away in the great flood of 1607.

We scramble to the top of the dunes and pick a route that keeps us on higher ground. I keep checking the map to make sure we are headed in the right direction. At the top of one of the tallest dunes, Al stops, looks around at the nature reserve below and says: 'I know this place. I've been here before. This is where we came to scatter my dad's ashes. He loved Kenfig so we brought him here to rest.'

I can understand. These dunes have an otherworldly feel about them. You sense spirits on the wind and a wild, free atmosphere orchestrated by the sea.

We finally locate the castle – now just a clump of derelict stone walls protruding out of the sand sitting under the elevated M4 motorway. The hum of constant traffic competes with the whistle of the wind. A herd of cows and a few wild horses observe us from a distance as the mist sweeps in again.

I walk up to one of the castle walls poking just a metre above the sand dunes. If the sand came that high up the castle, what have we just walked over to reach it? Have we been wandering on top of the old village, following the footsteps of a community that once called this home – only to see it taken from them by nature? It's a reminder of how quickly a landscape we assume is permanent can change. Of how the things we might take for granted today, may no longer exist in the decades to come. Most of all, it's a wake-up call about just how powerful the sea can be.

4

Copperopolis

Walks from Margam to Briton Ferry and a stroll around Swansea Bay

Mention Port Talbot to most people and a scenic coastal walk is not the first thing that springs to mind. This town near the mouth of the Afon Nedd, sitting across the water from Swansea, has long epitomised dirty, emissions-spewing heavy industry because of the enormous steelworks that dominates the town.

While the steelworks has long been seen as an eyesore by outsiders, it also has been the lifeblood of the town for more than 120 years. It was once capable of producing five million tonnes of steel each year and employed some 20,000 people. Today the steelworks' entire future is under threat. The coal blast furnaces were shut in 2024 with the loss of 2,500 jobs. Unless it can transition rapidly from its historic dependence on coal fuel to renewable energy – so-called green steel – it's likely that Port Talbot will become another of the ghosts of Wales's once mighty industrial past.

Copperopolis

Port Talbot's reputation also isn't enhanced by the fact that most people travelling by it have never really seen it. The M4 motorway cuts through the middle of the town, essentially dividing Port Talbot into two separate entities – the flat lower neighbourhood of terraced row houses built around the steelworks, and the hilly higher section notable for a number of large houses nestled on the side of Mynydd Emroch and that snake up the Afan valley. It's the sort of ruthlessly pragmatic urban transport project that Robert Moses, the infamous architect of New York City's highway system that cut through whole communities, would have undoubtedly been very proud of.

I want to get a different perspective. I want to see Port Talbot up close and it just so happens that the Wales Coast Path runs directly through the town. It has the potential to be one of the least glamorous walks in Wales. And I know just the people who will enjoy it. Andy and Jeff.

We begin on a dry but freezingly damp cold Sunday morning near Margam Abbey, once home to the Cistercian monks who lived and worked here in the Middle Ages. The town of Port Talbot didn't come into being until the 1830s when the wealthy Talbot family, owners of most of the low-lying land along the west Glamorgan coast, built a new dock at the mouth of the Afon Afan to handle the growing copper, coal and iron trade in the area. (The monks had first mined coal in the nearby hills nearly 500 years before.) As industry developed, the nearby villages of Margam, Aberafan (Aberavon in English) and Baglan became subsumed into the new town.

When I checked the map the day before, I was surprised to find that the Coast Path apparently offers two alternative routes through Port Talbot. One follows high ground starting in the community of Taibach (named after the small workers' houses that were built there in the early nineteenth century) before ascending through Pen y Cae and over Mynydd Dinas towards Baglan.

Seascape

I understand the appeal of keeping to higher ground. The hills above Margam and Port Talbot offer an immediate escape from the cramped feel of the lower town and provide walkers with beautiful views of Swansea Bay. Here, the trails over the hills take you north into the Afan Argoed and Pelenna forests that I walked through a few years before.

The high route will also be a smart alternate route if sea level and flood surge predictions for this part of the coast turn out to be true. The sobering projection is that, without robust and extensive new coastal defences, large parts of lower Port Talbot and Aberavon could be underwater within a generation. That gets me thinking again about what the Coast Path will look like, and how it will have to adapt as sea levels rise and flooding increases. Clearly new walking routes will have to be mapped and I make a mental note to be more aware on future walks of what routes might need rethinking. For today though, I'm set on following the route through the south of Port Talbot.

The path takes us through a mix of housing ranging from modest but neat to threadbare and neglected. We pass one patch of open fields with football pitches at the far end. A bright white sign at the entrance tells us this is a Bee Friendly zone. That's the one upbeat feel in this part of town. The rest feels robbed of optimism. Gulls swoop down over back lanes sprinkled with litter. The Coast Path sign is faded and decayed.

We meet two other walkers, a middle-aged couple from Hereford in England who are exploring the Coast Path over a number of months. They wear matching wet weather outer gear though the man sports a bright red cap while the woman opts for a blue bobble hat – perhaps to help tell each other apart. They are headed to Porthcawl, have left their car by Aberavon beachfront and are positively chipper about this section of the Coast Path. We are a little sceptical but no, they reassure us, we are in for a surprise.

They are right. Aberavon is a revelation – a nearly two-mile-long promenade and sands that look directly out at the Gower

Peninsula while also offering views east back towards Kenfig Sands and the Glamorgan Heritage Coast. Even the steelworks looks moderately appealing when viewed from here.

We stride along the prom, stopping regularly to take in just how good the view is. Neither Andy, Jeff, nor I – with a collected 150 years of Welsh walking experience among us – have ever visited this part of the coast. I know of Aberavon Beach because I've heard it's a good surfing beach. Andy points out this was where the Welsh rugby team used to come training back in the 1970s. Now I remember seeing TV footage of the greats – Gareth Edwards, Mervyn Davies, Gerald Davies and J. P. R. Williams – running and passing the ball on the sands.

We cut inland across Baglan Moor circumnavigating the old BP petrochemical facility that once sat at the mouth of Baglan Bay. The path tracks up the east side of the river to Briton Ferry where we can cross and continue on to Swansea.

Once this coastal estuary was the research playground where Alfred Russel Wallace, the famed naturalist and co-father of evolution, first fell in love with nature. It was here, over the river, in Crymlyn Bog and Burrows that he discovered the Phylan Dune beetle, more commonly known as the tiger beetle. That was in the early days of Wales's transformation into an industrial power and before this estuary was dredged into becoming Briton Ferry floating dock so it could handle the coal and other industrial goods being shipped to the coast on the Vale of Neath Railway.

Briton Ferry Dock was the brainchild of the prolific engineer Isambard Kingdom Brunel and is the only surviving example of his experimental work in the development of buoyant lock gates to control the water levels in a tidal estuary. Brunel also constructed a stone tower on the east bank of the river. It housed an Armstrong accumulator – a water-filled cylinder discharged by a heavily weighted plunger – that powered the hydraulic machinery in the floating dock. It was an impressive feat of engineering and another validation in the Victorian mind of how humanity could harness and control nature in the name of progress.

Today Briton Ferry's industrial power is just a memory and Brunel's old dock has long ago silted up – another reminder from the sea, perhaps, of who holds the real power.

We finished our walk that dreary day at the Burrows, on the outskirts of Swansea but, two months later, I have an opportunity to head back and explore the city's maritime quarter – another place that I haven't wandered before.

The chance comes in the form of an announcement by my daughter Zelda that she and her friend Mali have tickets to see the Arctic Monkeys on a Monday school night in Swansea. There is loose talk about them getting a train down and back from Cardiff but that would mean missing a chunk of the gig and, well, before you know it, I've volunteered to drive them and then hang around until th'Arctics have swaggered their stuff.

I project the air of dutiful father, making a minor though significant sacrifice to indulge my daughter's music education. Secretly I am quite happy to have four hours to explore the history of Swansea Bay. So, having dropped the girls off at the concert, I head down to SA1 – the renovated docklands quarter.

A metal statue of Dylan Thomas, Swansea's favourite literary son, sits in the square named after him at the entrance to the marina. I tip my hat, metaphorically, to the bard as I pass him and then the Wales National Waterfront Museum (making a mental note to come back and visit some other day). A small flotilla of pleasure craft bob in the marina next to it. In no time I'm at the boardwalk that stretches around the crescent bay to Oystermouth and the Mumbles, five miles away.

The sun is just starting its slow descent over the whitewashed Mumbles lighthouse far in the distance and the beach is busy with people enjoying the evening. A group of students huddle together against the breeze – drinking beer from cans and staring out to sea. A group of perhaps Syrian or Afghan families play down by the water's edge. The mothers stand in the surf, their hijabs fluttering in the wind and their dresses drenched up to the ankle, watching over young children who delight in being chased by the dribbling waves. The fathers stand further back up on the dry sand, delegating childcare so they can focus on smoking cigarettes and chatting. Further along the boardwalk, a West African family has fired up a portable barbecue on the grass outside the Swansea City Council headquarters. Another family – this one from eastern Europe – kicks a football around on the grass.

It's hard to imagine nowadays, when millions of us routinely head to the beach on sunny days to relax and swim, that, until the middle of the eighteenth century, most people in the British Isles avoided the sea completely unless sailing or fishing was their occupation.

It was around this time that the well-to-do began exploring southern Europe as the Grand Tour grew in popularity. In places like Italy's Amalfi Coast and Venice, they marvelled at how the locals swam in the sea… and appeared to enjoy it.

Yet even this cosmopolitan education wasn't enough to persuade them to take a dip in the decidedly cold waters surrounding the British Isles. However, the Grand Tour idea also arrived at a time when the genteel were just starting to understand the concept of health and wellness, countering an indulgent lifestyle that involved much rich food and too much alcohol (sound familiar?).

The solution advocated by the eighteenth-century intelligentsia was to take in the fresh air of the sea and to bathe in cold water – salty sea water in particular.

The trend towards embracing the beach first began in the 1720s. Bootle near Liverpool and Margate on the Kent coast were early destinations for those seeking a salt-water tonic. But it wasn't until 1750 that this new health craze started to become mainstream. That year, a Sussex doctor, Richard Russell, published a book titled: *A Dissertation on the Use of Sea Water in the Diseases of the Glands*. While Russell didn't specifically mention trips to the seaside, he encouraged the creation of spas where people could improve their health by bathing in, and even drinking, salt water.

His advocacy of sea water won him a place at London's Royal Society (the leading scientific academy) and started a new wellness trend that saw the development of spa facilities in inland towns such as Bath and Tunbridge Wells in England, and Llandrindod Wells in Wales.

As the idea of salt-water bathing grew in popularity, so people started bypassing the spa towns and headed to new beach destinations such as Scarborough and Brighton. In Wales, the Pembrokeshire town of Tenby got its first bathhouse in 1781 while, here in Swansea, bathing visitors started arriving from as far afield as the West Country in England.

The first part of the town to benefit from this new tourism boom was the Burrows, a strip of sand dunes and a long shallow beach near the mouth of the Afon Tawe. The Burrows, and the land directly around it, quickly became the town's fashionable quarter with new hotels and boarding houses opening to cater for the visiting bathing tourists.

By 1786, the *Gloucester Journal* declared that 'Swansea, in point of spirit, fashion, and politeness, has now become the Brighton of Wales.'

In 1789, to accommodate the new salt-water tourists, Swansea's town corporation agreed to build a dedicated Bathing House about a half a mile to the west of the town centre. The next year they gave the go-ahead for the construction of a bathing machine – basically a dressing room on wheels – modelled on the design being used at Weymouth on the English south coast. By 1798, Swansea had three machines for use by the guests who stayed at the Bathing House. They were operated by a driver for gentlemen (and a guide for the ladies). Both manoeuvred the bathers from the Bathing House down the beach into the sea but the guides helped the female bathers get into the water (such was the novelty of the experience). The most famous local female guide was Catherine Rosser, described at the time as 'principal operator and marine emerser to the fair maids of the Swansea flood'.

By the turn of the nineteenth century, Russell's ideas about sea water had become firmly embedded into the public consciousness. In 1803, a Swansea doctor called William Turton published *A Treatise on Hot and Cold Baths*, in which he extolled the benefits of salt water, 'which by stimulating the extreme absorbent vessels, excites them into more immediate action and produces a greater degree of that fine pleasurable glow upon the skin.' Today's Polar Bear all-weather sea swimming clubs would surely agree.

As these new seaside towns began to attract more visitors they discovered that bathing in machines wasn't enough to amuse people. These wealthy health seekers also wanted a bit of excitement so assembly rooms (where music and cards could be enjoyed), theatres, public gardens and piers were added.

Swansea boasted all these amenities including a pier which was completed in 1898 at the far western end of the bay at Oystermouth. But, by then, the town's destiny had been reshaped in a dramatic and brutal industrial way.

I walk further down the promenade, past where the original Bathing House once stood. Swansea rises to my right – its Uplands, Townhill and Mayhill neighbourhoods clinging to the high ground. Back in 1795, an English tourist raved about how the town was situated 'near the centre of a most beautiful bay, on an angle between two hills… making it a pleasant and very healthy situation.'

Yet it was exactly these geographical attributes that also made Swansea ideal for industrial expansion. The town was located on the western edge of the south Wales coalfield and the Tawe, which flowed into the sea at the Burrows, was deep enough that seagoing vessels could navigate as far inland as the mines.

That easy access to coal and the sea made Swansea a perfect place to make copper. In the eighteenth century, this metal transformed seafaring, and the British Navy in particular, through the invention of copper bottom hulls. The new ships were faster and nimbler than their wooden predecessors because barnacles couldn't attach themselves to them. In the 1760s, the British Navy ordered all its fleet to be protected with copper sheathing. By the mid-nineteenth century, copper also was becoming a critical component in the growth of mass communication through its use in the development of both the overland electric telegraph and sub-Atlantic cable.

Cornish and Devon copperworks had long dominated the industry. However they depended on imported Welsh coal for smelting and it took four tons of the black stuff to smelt a single ton of copper. Swansea offered an opportunity to cut costs. English copperworks owners began to shift production here and import the ore they needed from Cornwall, Devon and Ynys Môn's Mynydd Parys high above the town of Amlwch.

It was the opening in 1810 of the Hafod Copperworks, on the banks of the Tawe, that put Swansea on the world map. It was founded by a Cornish industrialist, John Vivian. At its peak it employed over 1,000 people. In 1835, a rival plant, the Morfa Works, opened near Hafod.

As demand grew, domestic copper ore deposits became so depleted that Swansea merchants had to search further afield. One port of call was Huelva Province in southern Spain. The workings around Las Minas de Riotinto are reputed to be the oldest in the world – having been in use for over 5,000 years with some saying they were the fabled mines of King Solomon.

Even they weren't enough to satisfy the appetite of the Welsh town now becoming known globally as Copperopolis.

Swansea ships also made regular trips to Santiago de Cuba near the Sierra Maestra copper mines and industrialists from the town established their own mining operations on the island. The vessels that sailed there took Welsh coal on the outgoing voyage to power the pumping engines of the mines, then returned laden with ore. In 1832, *The Cambrian* newspaper recorded the arrival of the *Emulous* from Cuba. It docked with a cargo of 200 tons of copper ore, eighty tons of rustic (a tropical hardwood tree) and 150 elephant tusks.

One of the furthest destinations for Swansea copper seamen was Chile. These hardy folk were known as Cape Horners and respected throughout the town for their skills in navigating the Southern Ocean to bring back ore from the Pacific port city of Valparaíso. By 1866 more than £1 million worth of copper ore a year was being exported to Swansea from mines in South America and Chile in particular.

At the height of the industry, thirteen copperworks were going full blast in the Swansea area. Copper brought wealth to the district, but at a substantial cost. The process of separating the metal from the ore produced mountains of slag and furnace ash along with billowing clouds of foul-smelling smoke laced with sulphur and arsenic.

Social status soon became determined by how far away you could live from the copperworks. While at the end of the eighteenth century, the well-off had settled near the water's edge, now they chose to live on the hills overlooking the town while the copper workers and their families lived in row houses hastily built near the plants on the banks of the Tawe.

In the 1770s, Swansea had a population of just 1,500 people. That had swelled to over 10,000 by 1821 and most of them were crammed into squalid neighbourhoods next to the copperworks.

Living conditions were appalling and the workers and their families were known for their sallow complexion and the afflictions they suffered – principally tuberculosis (known then as consumption). According to the historian W. R. Lambert, 'when men from outside the district were hired by the copper works, they either became used to the conditions or died within a very few months.'

The town found itself at a crossroads – how could it maintain its growing reputation as a seaside resort while also developing as a globally important copper producer? It was a concern for the wealthy investors in Swansea's tourism trade not least because they were some of the same people who were driving the town's industrial growth.

In 1822, Hafod's owner John Vivian and fellow industrialist Lewis Weston Dillwyn, owner of the Cambrian Pottery Company, proposed a £2,000 reward to anyone who 'in the most satisfactory manner, shall destroy the noxious qualities upon the whole process, and, at the same time, effectuate the greatest reduction of the bituminous smoke, upon the plan adapted to the present practical operations of copper smelting and at the most reasonable expense.'

They didn't know it at the time but, essentially, these two Victorian entrepreneurs were calling for the invention of carbon capture and storage – the same technology that coal, oil and steel producers today promise (with their fingers crossed behind their backs) will help them operate without further contributing to climate change.

The reward was a sizeable sum and the industrialists – both being fellows at the Royal Society – were able to seek counsel from some of the greatest scientific minds of the day including Michael Faraday, whose breakthrough studies into electricity and electromagnetism would soon transform the Victorian world.

We don't know whether Faraday ever took up the challenge of ridding Swansea of its noxious copper fumes but he was certainly aware of the problem.

'The extensive copper works towards Swansea sent up an emmense cloud of sulphurous smoke, which … obscured and hid the scenery,' Faraday observed on one visit in 1819.

Ultimately, while seaside tourism continued to grow to the west in Oystermouth and the Gower Peninsula, the industrialists chose copper over sea bathing. By the 1840s, a new dock was needed to handle the volume of ships and cargo entering and leaving Swansea. The decision was made to build those docks in the Burrows, once considered the most fashionable place to live.

It's almost time to pick up my daughter from the concert so I wander back down the main road towards the marina. Here, just inland from the site of the old port, sits a new architectural homage to Copperopolis. Known as Copr Bay, it features the first new public park to be built in the city since Victorian times and it houses the Swansea Arena auditorium – its Frank Gehryesque façade shimmers copper. The metal railings protecting the mezzanine walkway also are painted copper as is the footbridge over Oystermouth Road featuring silhouettes of swans on its flanks. The bridge was designed by an artist friend of mine and Swansea native, Marc Rees, to acknowledge and honour the importance of the metal in shaping his hometown, and as a physical way of reconnecting the old industrial downtown district to the beach and sea.

I walk to the middle of Marc's bridge and watch the traffic roll by underneath. Two young women cross the street below. They are clearly dressed up for a night out on Wind Street, Swansea's drinking quarter – they teeter somewhat in their heels. A sports car full of young men speeds by and honks its horn. The girls turn and respond with an emphatic two finger salute.

Just out of view is the local Sainsbury's superstore. This was where, at the height of the copper boom, The Cuba Inn once stood – a legendary pub where sailors would gather for a final drink and a bit of rowdiness before heading off across the seas. It seems that, while the pub and the industry may have gone, the old docks' sensibility still endures.

There's one more place in historic industrial Swansea that I want to see so I walk a short distance inland from Copr Bay through streets of row houses until I reach the police station. There, on the exterior wall is a blue plaque celebrating the life of an inventor, William Robert Grove, who grew up in a house on this site. You may never have heard of him, but his most famous invention – the hydrogen fuel cell battery – will transform how we live in the twenty-first century.

Grove was born in 1811 to a wealthy local family. His father was John Grove, a magistrate and deputy lieutenant of Glamorgan, who ran in the same circles as the Vivians and Dillwyns. He left Wales to study at Brasenose College, Oxford before qualifying as a lawyer. But Grove's real passion was science and, while at university, he became a member of the Royal Institution, a new society for the leading scientists of the day.

In 1837 Grove gave a talk at the Institution describing 'an economical battery ... made of alternate plates of iron and thin wood, such as that used by hatters'. He would expand on his thinking in a scientific paper titled 'On a New Voltaic Combination' that was published in the *Philosophical Magazine and Journal*.

This early work on what we now know and take for granted as battery technology would prove instrumental in helping power the growth of the American telegraph system. This and other quite spectacular thinking – including the invention of an incandescent electric light that would lead to the evolution of the lightbulb – soon got the attention of Michael Faraday.

It also piqued the interest of English inventor Charles Wheatstone, who was conducting his own experiments with underwater electronic telegraphy and would use Swansea Bay as the testing ground for his first submarine communication experiments. Wheatstone was so impressed by Grove that he put him forward for Fellowship of the Royal Society.

In October 1842 Grove corresponded with Faraday about another new breakthrough. He described how his invention involved dipping platinum foil into alternative tubs of oxygen and hydrogen. The result, he said, was 'an unpleasant shock...'

Grove had invented the hydrogen gas fuel cell – a zero-carbon way of generating electricity.

He and his contemporaries clearly envisaged a new modern world powered and enabled by electricity. To a great degree they were correct, though none of them in the 1840s could have anticipated the way a new fossil fuel, oil, would soon start to dominate power generation. Today though, as the era of oil begins to dwindle and the rush to replace its potency with renewable energy intensifies, Grove's invention is more important than ever.

The sun has set and the chill from the bay has intensified so I retrace my way to the marina. The families and students have left, but their footprints, and those of the others who walked here this evening, remain in the thick dusting of sand that has swept over the tarmac Coast Path. I look back across the bay towards Port Talbot and the imposing shell of the steelworks. Will it still be operating in ten years' time? Only, perhaps, if the steel makers successfully make the switch to hydrogen fuel to create green steel. If that happens, it will be the local boy William Robert Grove who they can thank.

Putting Faith in Science

Walks from Rotherslade to Rhossili and Llanmadoc to Burry Port

In August 1897 the Anglo-French Impressionist artist Alfred Sisley arrived on the Gower Peninsula as part of his honeymoon celebrations. He and his long-time partner Marie Lescouezec had just married in Cardiff and were staying in the Osborne Hotel, overlooking Little Langland – or Ladies' Bay as it was known at the time because it was a place where women could swim separately from men.

Sisley was almost destitute having failed to achieve the level of acclaim enjoyed by his friends Claude Monet, Edgar Degas and Pierre-Auguste Renoir. He and Marie had two grown children by the time they got married and were only able to make the trip thanks to the largess of a patron, François Depeaux, who appreciated Sisley's understated, subdued approach to painting. Here, at Little Langland (now better known as Rotherslade), Sisley would tap into that artistic inspiration and capture the south Wales coast in a way few had done before.

A critic in the French paper *Le Journal* later observed that Sisley had brought back from Langland Bay 'a series of admirable sea pieces, in which the strange flavour of that landscape, little frequented by painters, is rendered with an art that is as captivating as it is personal.'

One of the main muses for Sisley was Storr Rock (known in Victorian times as Donkey Rock), the tall, rounded mound that dominates the top of Rotherslade beach. Sisley painted the rock and other views of Little Langland and greater Langland Bay to accompany the seascapes of Penarth, Flat Holm and Steep Holm that he had finished earlier in the month. In October he and Marie returned to France but both died within two years of being wed. It was only after Sisley's death that his work began to gain respect with critics and collectors. Today, his painting *Storr Rock, Lady's Cove, le soir,* is part of the National Museum of Wales's permanent Impressionist collection.

I knew nothing of Alfred Sisley when I was kid but I knew the rock better than any other part of the Welsh coast. Rotherslade was my family's favourite beach to visit when I was ten years old. It was about an hour's drive from our home in Cardiff and, importantly, was accessible via an old two-storey promenade with steps down to the beach. That meant my grandmother (and her creaky knees) could join us for the day. On one floor of the promenade was an ice cream shop where I'd queue for a treat.

Reaching the summit of the rock was a rite of passage for kids of a certain age (and height). Really small kids didn't stand a chance of summitting while teenagers found the endeavour a little too easy to brag about (even though they did it anyway). But if you were eight, nine or perhaps even a diminutive, skinny but optimistic ten-year-old, getting to the top was an ultimate ambition – especially at high tide when you could marvel at the big kids jumping off the rock into the sea. I must have tried for what felt like almost an entire summer to hoist myself up and over one section of the rock before finally succeeding. My knees were scraped and battered by digging into the barnacled rockface for traction but I made it. Even today, nearly fifty years on, I can still recall the thrill of finally reaching the top.

It is with those memories fresh in my mind that I return to Rotherslade this spring morning. The old rusting promenade, with its stairs that always smelled of a combination of sea water and urine, has long since been demolished. It had been built in the 1920s to help shore up the eroding cliff face. In its place is a paved walkway, with a café and tables and a far more sanitary descent to the beach.

The tide is out so I walk down over the first barrier of large grey pebbles until I reach the firm, dark yellow sand that made Rotherslade the perfect place for beach cricket when I was young. This is also where I asked Jowa (in a howling gale on Boxing Day) to marry me twenty-five years ago. It's where we brought our own kids over many summers and where I watched them conquer Storr Rock just as I had decades before.

My plan is to take a walk back in time to Oxwich Bay some ten miles away – revisiting parts of Gower that meant so much to me growing up. Long strips of High Tor Limestone and Hunts Bay Oolite rock extend down into the sand on the approach to Langland Bay beach. It was here, in Tor Cave in 1892, that workers discovered a Mesolithic mammoth's tooth when they were building the Osborne Hotel. Being pragmatic, they quickly filled and blocked up the cave to secure the hotel's foundations.

The next bay after Langland is Caswell – a favourite destination for families and surfers. The tide has retreated for the morning and the stream that runs down the middle of the beach sparkles as the sun reflects off the shallow waters. The Coast Path offers two options here. Take the high tide route in front of the grand houses that overlook the bay or the low tide walk across the beach? I choose the latter – eager to get as close to the sand and the sea as possible and lose myself in its sounds and smells.

A group of children has gathered halfway down the beach, listening intently to a young woman who turns out to be their teacher. The kids are from the local primary school and this is one of their regular beach lessons – a chance to study in nature. Right now, they are having a maths lesson – doing calculations in the sand. Next, they will be learning about Celtic knots and ancient Welsh hill forts, their teacher tells me. 'We are just so lucky to have the beach on our doorstep and make it both a classroom and a playground,' she says.

A thin vapour of sea water lingers above the hard, wrinkled sand as I walk to the edge of the beach. I look back at the children running around, picking up bits of washed-up wood and digging in the sand. It makes me smile and also think about how much focus now is being made in Welsh schools on developing an understanding and appreciation of our history and culture. It wasn't always that way. Especially not here, in this anglicised area of the south Gower which, in some ways, has more in common with parts of Devon and Somerset than Wales.

Most of south Wales is predominantly English speaking but that's mainly a result of industrialisation over the past 200 years. Here though English, and a particular dialect of it, has been the dominant language since Norman times. From the early twelfth century, Gower became one of the Welsh regions under Norman Marcher Lords' control. The peninsula was divided into two areas: Gower Anglicana (English Gower) in the south and west, and Gower Wallicana (Welsh Gower) in the northeast. English-speaking peasants from the West of England were persuaded to settle in Gower Anglicana, perhaps because the Welsh inhabitants revolted against the Marcher Lords. Over generations, this southern part established its own linguistic and cultural identity.

The new dialect was reinforced by regular trading across the Bristol Channel so that, until the growth of industrialisation in the nineteenth century, the grammar and pronunciation of the Gower dialect was very similar to that in Devon, Somerset, Dorset and Cornwall.

It was the copper industry in Swansea that changed the sound of south Gower – ironically because a new generation of English-speaking immigrants came to live in south Wales. The Welsh linguistic boundary across the north Gower was slowly eroded by the new arrivals and, with English becoming the dominant tongue throughout the region, the English Gower dialect was lost.

Of all the Welsh myths surrounding the castles and villages swallowed by the encroaching coast, I think the story of Pennard Castle's demise might be my favourite. About three hundred years ago, Pennard also started to succumb to the dunes. Here, legend maintains that the catastrophic event was the result of a curse put on the Lord of Pennard by local fairies – payback for threatening them with a sword when they danced and frolicked in the moonlight.

The ruins of Pennard Castle lie close to Three Cliffs Bay and a little beach known to everyone locally as Pobbles. Decades ago, a family friend, Ishbel, lived nearby and so the walk from the village of Southgate down to the beach was one I made countless times as a child. When I was eight years old, this path signalled the start of what seemed like an endless hike over cliffs and across the sand dunes to Pobbles. It was particularly arduous for little legs when you had to carry a cricket bat, a football, a bucket and spade and your own towel.

I buy myself a coffee at the local shop in Southgate and start retracing the walk I did as a child. It's not nearly as hard work as I remember it. I stride down the paved road passing fancy new-looking holiday homes that didn't exist forty years ago. One of them has tall metal gates to restrict the prying eyes of people like me. On another house, two gulls stand sentry equidistant from each other at opposite ends of the roof. A solitary red kite hovers over the proceedings – looking out for quality pickings from the rubbish bins no doubt.

Tarmac gives way to grassland which then frays and disintegrates into soft sand as the path starts a steep descent down to Pobbles. Over on the next bluff, cows graze on what looks like the fifth fairway of Pennard Golf Club.

At Pobbles beach, as I step over the top layer of pebbles and start to walk on the sand, I'm overwhelmed by memories of my youth. Of how the thin outer crust of the hardened sand crumbles with each step you take and how my feet sink into the soft sublayer below. Now I remember why Pobbles was not a great beach to play cricket on.

The sand hardens as it gets wetter and I approach the wide but shallow stream that runs down the middle of the adjacent Three Cliffs Bay. It presents me with a dilemma. Shall I remove my boots and wade, or back myself to spot a route across the stream that isn't too deep?

Going barefoot is clearly the most pragmatic option but I don't have a towel and am not keen to put socks back on wet, sandy feet. So I keep my boots on – only to fall knee deep into the current within thirty seconds of attempting to cross the stream.

I stop, sit on a big rock next to a tide pool, curse my stupidity and laziness and slowly pull off my soaking wet boots and socks. It is low tide and the beach stretches all the way from here to Oxwich – my end point for the day – at the far end of the bay.

I trudge barefoot past Great Tor, a grisly, weathered slab of granite that has been here since the Ice Age. It looks like a miniature version of Hallgrímskirkja, the iconic Lutheran church in the heart of Reykjavik, its vertical layers of light grey rock reflecting the sun. For the next thirty minutes, I pick my way through razor clam shells and sharp, tiny beach stones – all the while my feet getting colder and colder. The waves break on the beach in what sounds like white noise chatter. I suspect it's the sea having a chuckle at my expense.

Towards the end of 1822, two brothers, the Reverend John Davies and Daniel, a surgeon, decided to explore a coastal cave known as Goat's Hole, two miles west of Port Eynon, roughly halfway along the southern part of the Gower coast. In it they discovered some Roman coins. Their interest piqued, they took the coins to the Mansel Talbot family, owners of nearby Penrice Castle, located a couple of miles inland from Oxwich Bay, so that they could be added to a private museum housed there. The Mansel Talbots – the industrialist founders of Port Talbot and a core part of the Swansea aristocracy – also were intrigued – particularly Lady Mary Talbot. They sent word of the find to Lewis Weston Dillwyn – the owner of Cambrian Pottery whom they knew had important connections at the Royal Society and the Royal Institution.

Dillwyn immediately headed to Penrice. Over the next few days, he, Lady Mary and another guest at the estate, travelled the five miles overland to explore Goat's Hole cave. There they discovered something quite remarkable, though nothing to do with Roman coins. In the cave was a set of bones including a mammoth skull and tusks.

The group didn't fully understand the importance of their find, but they knew someone who might. Both Dillwyn and Lady Mary were acquaintances of the Reverend William Buckland, the famous professor of Geology at Oxford University. In early January 1823, Miss Talbot wrote to Buckland. He wasted little time heading to the Gower coast and spent a few days exploring Goat's Hole and excavating more animal skeletons.

It was at some point during his visits to the cave that Buckland made an amazing discovery – an undisturbed burial of human bones stained red with ochre. There was no skull but the bones were surrounded by ivory objects (including rods and rings), periwinkle shells and worked flints. Buckland collected the remains and returned to Oxford.

It was a time of great breakthroughs in the world's understanding of geology and the physical age of our planet. Buckland was aware of, and somewhat open to, new scientific theories but he was also absolute in his Christian belief that no human remains could have existed before the time of the Biblical Great Flood sent by God to rid the world of evil and, for want of a better phrase, reboot humanity.

Governed by this doctrine, Buckland first posited that the bones must have been those of a male customs tax collector who had been murdered (presumably by smugglers who used the caves to store their contraband). Then he postulated that the remains were female and belonged to a witch, due to the presence of a 'blade bone of mutton' – apparently interpreted as a tool to conjure the spirits. Finally, he settled on the lurid but no doubt attention-grabbing theory that the skeleton was a female prostitute.

Over time, his find became known as the Red Lady of Paviland. But in being so absolute in his desire to marry science and his religious beliefs, Buckland failed to comprehend the skeleton's true significance. The Red Lady was really a young man who had been buried in the cave some 30,000 years ago when this part of the coast was situated seventy miles inland from the nearest ocean. Paviland, as the cave is now known, turned out to be the oldest ceremonial burial place in Western Europe, dating back to 20,000 years before Stonehenge was built.

My walk today is taking me up the Coast Path from Oxwich and Penrice Castle to Rhossili, one of the longest and most popular beaches on Gower. I've already passed through the villages of Horton and Port Eynon and now I've reached the location of Buckland's great discovery. There's a weathered wooden sign

Seascape

pointing out the cave. Standing at the top of the cliff looking down towards Paviland nearly seventy metres below, I see a dusty single step track, punctuated by boulders, offering the chance to descend. Someone has run a length of rope part of the way to help negotiate what looks like a tricky section.

I could ignore this detour and keep walking. But how often do you get to walk back in time? I scramble down the track, stepping carefully over and around boulders until I reach the tiny inlet that leads out to the cave. Every rock in the inlet has been buffed baby skin smooth by the waves.

This particular day the cave itself is out of reach. The tide has turned and is lapping at the mouth of the inlet where I'm sitting. As I watch the sea coming in (quite rapidly), I try to imagine what life might have been like for the young man who lived here some 30,000 years ago. What would he have seen from this vantage point? It certainly would have been very different from my view today. Back then, this inlet would have been high ground looking across a fertile plain stretching for miles below. Before I have time to contemplate any further, a wave pushes through and reaches my feet. I scramble back out of the inlet and head for higher ground.

A few weeks later, I return to Gower with Jeff, Al and another friend, Tim, to walk from Llanmadoc to Penclawdd. Today's tramp will mostly involve navigating the salt marsh – a constantly changing landscape of mudflats and tidal ditches – that dominates the north coast of this peninsula. It's not a section of Gower any of us knows well. There's an air of expectation even before we embark on the ten-mile journey ahead of us.

We leave the village of Llanmadoc and follow a lane towards the coast. Ahead of us, the barks of the trees are bleached white, the branches brittle, the leaves no more. We have stumbled on a ghost wood – a cluster of tall but ghoulish trunks that stand at the northern tip of Gower, where Broughton and Whiteford Burrows converge. These burrows are internationally recognised as a feeding ground for wading birds and wildfowl including oystercatchers, knots, pintails and golden plovers.

We follow a coastal path on the OS map that leads towards Whiteford Point – a spit of low-lying sandy land spreading out towards Burry Pill in the Afon Llwchwr (Loughor) Estuary. Tall yellow loosestrife grow out of the sandy soil on the fringe of the salt marsh. To my untrained eye, they resemble giant mutant daffodils.

I'm searching for a track that will provide a short cut across Landimore marsh. I've seen it on the OS map and the route looks promising. We join it as it leads across a long berm-like causeway that has been built to keep the tide at bay – the marshy salt flats occupy the ground on either side. Within minutes, though, we spot a problem. Ahead of us, the causeway has been breached. Destroyed might be a more accurate term.

The only way to cross the gaping hole is to descend into the salt marsh mud. Al takes the lead – showing that impish enthusiasm he brings to every walk. He skips his way down off the berm, finding odd bits of wood and stones to stand on so he can piece together a path across the mud without getting stuck. Tim and I go next, happy to follow in his precise footsteps. Jeff seems less keen. He is an excellent walker as long as he's going in a straight line but blown ligaments in both knees make bending and twisting potentially calamitous. By the time we have made it across the soggy salt marsh, Jeff is only just easing himself down from the berm.

The truth is, had I done better research before we set out this morning, I would have known that this sea wall has been broken for nearly a decade.

The Cwm Ivy marsh had first been reclaimed from the sea in the seventeenth century. Farmers protected these arable lands with a wall which over the years increased in size and strength (today it is the berm we've been walking along). Then, in late 2013, the wall started showing signs of distress. The pressure of water had forced a small hole under it. The following winter, rain, high tides and storm surges began to widen the hole and allow significant amounts of sea water into the freshwater marsh.

The sluice gate designed to drain the marsh couldn't remove the water fast enough and in 2014 part of the wall collapsed. The farmland grasses died within days and the trees rapidly began dropping leaves. That accounts for the 'ghost' woodland we just passed. The salt water had poisoned the trees from the roots up.

The following year, the National Trust, owners of this land, made the decision to surrender Cwm Ivy once and for all to the sea. Part of the rationale was based on minimising 'coastal squeeze', a term used to describe the damaging effect when natural habitats can become trapped between rising sea levels on one side and infrastructure such as sea walls, roads and natural features like hills on the other.

It had published a report titled *Shifting Shores* that highlighted how parts of Wales could no longer build their way out of coastal erosion and the impacts of climate change. 'This means at sites such as Cwm Ivy, we no longer try to defy nature by holding back the tide, instead we let nature take its course,' it wrote. Since then, the cultivated farmland has returned to salt marsh, joining the great expanse of *morfa* that spreads along the Llwchwr Estuary.

Tim, Al and I make it back up to the top of sea wall and stop to witness the spectacle of Jeff picking his way tentatively across the muddy marsh before painstakingly climbing back up to us, which he does with a certain amount of huffing, puffing and cussing.

'I'm not going to blame you quite yet,' he says to me once he reaches the top. 'But I suspect this will prove to be yet another of your ill-considered short cuts.'

We resume our walk along a potholed single-track road that cuts through the salt marsh in more or less a straight line up the estuary. Baby lambs wander the *morfa* while large orange and brown marsh fritillary butterflies flutter around us. In the distance lies the village of Crofty. Along with its next-door neighbour, Penclawdd, Crofty is the centre of Welsh cockle harvesting – an industry that in the middle of the nineteenth century employed people all along the wide and shallow sandy estuaries of the Welsh coast and sated the appetites of shellfish lovers throughout Wales and as far afield as the English Midlands.

At low tide, this estuary is a wealth of sand and mudflats where cockles thrive. They are some of the most sought after in the UK and there's evidence that cockles have been gathered along this coast since Mesolithic times. The Romans harvested them on a commercial scale when they established a fort in nearby Leucarum – what we now know as Casllwchwr (Loughor).

One report from 1916 estimated that almost 320 tonnes of cockles were harvested in the Penclawdd area each month. On one typical day visiting officials from the South Wales Sea Fisheries Association observed more than fifty women working on the beaches. Each of their donkeys carried some 150 kg of cockles in sacks.

This truly was women's work. Generations of them learned the changing of the tides and where the best parts of the sand were to rake for cockles while their husbands mostly worked in the region's coal mines or in the local copperworks that John Vivian had built in Penclawdd.

It might seem strange today but just a generation or two ago, cockles were a popular snack and cooking ingredient throughout Wales – sold at local fishmongers or even by mobile vans. Generations before, women would sell them by going door to door – carrying wooden pales of boiled and shelled cockles on their head and uncooked ones in baskets under their arms.

The north Gower cockle industry has shrunk considerably over the past fifty years – the victim of changing consumer tastes that have been conditioned to eat ultra-processed fast food rather than snack on fresh seafood. The cockle population also suffered from an invasion of parasites in the early 2000s that resulted in mass mortalities. It, in turn, caused a decline in the wildlife that fed off the cockles – notably the local oystercatcher population.

Despite the many setbacks, the cockle industry along the Llwchwr Estuary endures partly because of regulations that established a quota on the number of cockle licences that can be issued, a stipulation that all raking must be conducted by hand and a quota on how much can be taken each day. As a result, the companies that remain in Penclawdd are part of the only major cockle industry in the UK that still uses hand raking, although they now rely on tractors rather than donkeys to carry the cockles off the beach. Most of their harvest is exported to markets in Spain and elsewhere in Europe where cockles remain a delicacy.

Al and Tim have gone back to Cardiff so it's just Jeff and me this morning, exploring the north stretch of the Llwchwr. Ahead of us lies the town of Llanelli, famous throughout the sporting world for its rugby team. But before we get there there's a little detour I want to take Jeff on.

Not far from the bridge that allows the Coast Path to traverse the Afon Llwchwr sits the town of Casllwchwr. Roman strategic pedigree aside, it hasn't been much of a must-visit destination since. Apart, that is, from a couple of years at the turn of the twentieth century when, for a fleeting moment, all the eyes of the world focused on the miraculous goings-on in this small town.

Back in 1904, a young man named Evan Roberts returned home here on a spiritual mission. He was convinced that the Holy Spirit wanted him to spread the word of God to the young people of his local Moriah Chapel.

Roberts was just twenty-six at the time and had been away studying to prepare for the priesthood on the west Wales coast in Ceredigion. He began running a series of prayer meetings at

Moriah where he urged the young people present to rise and confess their commitment to Christ publicly. Only seventeen people attended that first meeting. Many had just turned up out of curiosity but something about his passion and fervour energised the congregation.

Soon Roberts was leading a full-blown religious revival that spread rapidly through both industrial and rural Wales. Attendees marvelled at how the Holy Spirit spoke through the tongues of some enraptured congregation. He began touring throughout Wales holding Revival meetings that would last all evening and into the early hours of the morning so that coal miners ending the night shift could attend.

Moriah Chapel is located about a mile inland from the Coast Path. Surrounded by a new-build housing development, its austere brown-stone façade feels stranded in time. There is a memorial stone to Evan Roberts outside the red front doors of the chapel but I do wonder if most local people even know of his influence over Wales and the rest of the world.

Within six months, Roberts had converted more than 100,000 people. It was said that drunkenness, violence and gambling declined and that groups of miners prayed together before heading down the pit for their shift. Evan Roberts's Welsh Revival was as potent as any of the non-conformist religious awakenings that had taken place for the past 200 years in Wales. But while his fervour burned bright, it would be extinguished within just two years as the pressure of near constant proselytising affected his well-being. Soon, he was doubting the 'voices' he was hearing in his head, he suffered a severe mental breakdown and withdrew from public meetings. He left Wales to convalesce in England and didn't return until 1930. Roberts died in 1951 and is buried back where his religious journey began: at Moriah Chapel.

That might have been the end of the story were it not for the fact that Roberts's Welsh Revival movement had attracted the eyes of the world. Within a year of his preaching, word spread through newspaper reports and letters to Europe, Africa, Patagonia, China and Korea where missionaries promoted the Revival. But it was in the United States that his words had the most impact.

One of the people who Roberts corresponded with during 1905 was a Los Angeles-based journalist named Frank Bartleman. He had read about Roberts and had written a letter to him requesting he 'pray for us in California'.

The two men struck up a correspondence and Bartleman spread the word of Revival through newspaper columns and talks at local Los Angeles churches. One of those was on Azusa Street. There, on 18 April 1906, the *Los Angeles Times* reported on a new type of religious gathering where the congregation communicated in a 'weird babel of tongues'. It became the foundation of what we now know as the modern Pentecostal movement – embraced by more than 656 million evangelical Christians worldwide.

We rejoin the Coast Path and soon reach Llanelli, the largest town along this part of the Llwchwr – once famed throughout south Wales as a centre not just for coal exporting, copper smelting and iron works but also for its zinc and tinplate industry. At its industrial height in the 1890s, twenty tinplate works operated around Llanelli accounting for some ninety per cent of the world's supply. That made it an industrial rival to nearby Swansea. You won't be surprised to know that the locals refer to the town as Tinopolis.

The Coast Path takes us near a neighbourhood called Seaside on our way to the town's revamped beachfront. A historical tourism sign tells us that this district grew up as a result of the shipping trade along this section of the Llanelli foreshore known as 'The Flats'. Two iron and copper entrepreneurs named Alexander Raby and Charles Nevill created Seaside – essentially reclaiming the mudflats by dumping copper slag so that the foundations of the new neighbourhood buildings sat above the high-water mark of the estuary.

At the mouth of the Afon Lliedi, Jeff and I stop for a coffee at a beachfront café before strolling towards Burry Port. This part of the fourteen-mile-long Millennium Path – a grand scheme that transformed 1,000 hectares of decaying docklands and derelict industry into coastal parkland – is like cruising down a US interstate highway compared to most of what I've encountered on my walks so far. The wide, winding paved path has been constructed very much with cyclists in mind so they can ride down to Pembrey Sands and back.

Shallow pools of water pockmark the Cefn Padrig sands that stretch far out into the bay at low tide. On the far coast Llanmadoc Hill on north Gower towers above the salt marsh while, at the tip of the peninsula, the tall sand dunes at Broughton Burrows look like breaking waves from this distance.

Outside the village of Pwll, we pause to read a blue plaque that reads:

Amelia Earhart
1897–1937
First woman to fly across the Atlantic Ocean
Landed here in the estuary near the village of Pwll
18 June 1928
Sponsored by the Pwll Action Committee

Amelia Earhart was an assistant pilot on a truly landmark flight that saw her fly from Newfoundland across the Atlantic to the UK. The orange Fokker F7 seaplane, called the *Friendship*, was scheduled to land in Southampton but had run low on fuel and so found refuge in the Llwchwr Estuary. Earhart didn't fly the plane – her two co-pilots took the controls for the twenty-hour journey – but her achievement of being the first woman to fly across the Atlantic would go down in history.

The little village of Pwll suddenly had a claim to fame. Except things weren't that simple because, having landed in the estuary, the plane was then towed into harbour at nearby Burry Port where the local people made sure to entertain Earhart for the evening. Later Burry Port would claim her visit as part of their own town's achievements.

Ahead of us is Burry Port lighthouse, positioned near the entrance to the harbour. Opposite it, on the very far tip of Gower, sunlight bounces off the cast iron façade of Whiteford Point lighthouse.

There's a very good reason why this part of the estuary boasts not one, but two lighthouses – it's a particularly dangerous area of coastline. The beaches around Burry Port – notably Cefn Sidan to its west – are littered with the wooden graves of ships that ran aground on the hidden sandbanks just offshore. Many vessels were said to have been lured to their fate by teams of 'wreckers' who lit lanterns to confuse those navigating and draw them onto shore.

Such was the reputation of these local villagers that they gained a nickname '*Gwyr-y-Bwelli Bach*' (the men of little hatchets) because of the clawed tool they used to prise open contraband, remove jewellery from the bodies and fingers of the drowned sailors and, allegedly, dispatch those unlucky enough still to be alive once the wreckers arrived at the scene.

It's impossible to say how true these lurid tales of ambush and murder might be. Certainly, a great deal of looting went on. In 1818 *La Providence*, sailing from Bordeaux to Dunkirk, ran aground on Cefn Sidan laden with a cargo of juniper berries, wine, brandy and coffee. The authorities found many local people on the beach completely intoxicated. One man was said to have died from drinking so much.

Ten years later *La Jeune Emma* was sailing from Martinique to Le Havre in France with rum, sugar and coffee when it lost its bearings and ended up wrecking on the sands. Thirteen people died including Lieutenant Colonel Coquelin and his daughter Adeline, niece of Josephine, consort of Napoleon Bonaparte. It was reported that Adeline was missing two fingers when her body was found – a victim of the '*Lladron Glan y Môr*' (robbers of the sea) according to reports at the time.

Putting Faith in Science

We stop at Burry Port harbour to soak up the views across the estuary and take stock of where we've walked. During the last few days on this small stretch of the Welsh coast, not only have I been able to revisit seaside memories of my youth, I've also delved into one of the greatest scientific discoveries about human existence and learned about one of the greatest displays of, and influence on, religious faith.

Two hundred years ago the Reverend Buckland wasn't able to countenance that his discovery of the mysterious Red Lady could be older and hence more important than his religious worldview allowed. This was despite a growing acceptance within academic circles that science, not God's Will, was a more accurate way of explaining the physical history of the earth. This was an awakening that Daniel Defoe had started to embrace just over a century before (even though he still believed that the Great Storm of 1703 was due to divine intervention). And it would explode into mainstream consciousness a few decades later with Alfred Russel Wallace and Charles Darwin's theories of evolution.

Today our society is still constantly trying to reconcile the reasoned findings of science with the absolute faith of people's convictions just as the Reverend Buckland did – no more so than in the ways we react to climate change.

There is near unanimous acceptance in the scientific community that climate change is driven by humanity's use of fossil fuels. Much of the non-scientific world also now agrees and supports (in principle) the superhuman task of transitioning away from fossil fuels to renewable energy. The reality, though, is that any transition will be economically painful for sections of society. Many people (not least those with vested interests in the energy industry) continue to argue that fossil fuels will be needed for many decades to come, even if that means exacerbating the devastation caused by climate change.

Further muddying the debate is a growing distrust or plain disbelief in the factual findings of science – part of a 'post truth' movement that has been fuelled over the past two decades by the growth of social media and certain influential voices (many funded, once again, by vested interests).

Is it any wonder then that many sections of society would prefer to trust their own faith in the power of their god rather than 'findings' of scientists? In 2022 the Pew Research Center surveyed American opinions towards climate change based on religious beliefs. It found that evangelical Protestants were the most likely of all major US religious groups to express sceptical views. Notably, nearly a third of Evangelicals said that climate change is not a serious problem because 'God is in control of the climate'.

If your religious convictions – the principles that guide how you live a good life and which you believe help create a better humanity – maintain that your god alone has control over our world's climate, that's quite difficult to reconcile with the need to overhaul our energy infrastructure and economic systems.

So far, the argument that decarbonising our economy is the only way to tackle the climate emergency hasn't been won. Instead, Evan Roberts's wholehearted embrace of 'God's will' continues to resonate in ways that influence our futures – including those living on this particularly vulnerable estuary.

6

Below the Landsker Line

Walks around Laugharne, from Amroth to Tenby and Freshwater West to Angle

I've come to Laugharne to speak at a small, well respected literary and music festival, The Laugharne Weekend. My talk is on the Friday night which gives me all of Saturday to explore the village and the festival.

I'm staying five miles outside of Laugharne in the village of Llanddowror – such is the popularity of the festival that most accommodation fills up quickly. So, over breakfast on that Saturday morning, I research the best walking route down into the coast town. Not more than half a mile from my hotel, there's a public footpath that cuts diagonally through the fields and country lanes until it meets up with the Afon Taf.

Walking into Laugharne interests me for a couple of reasons. First, I love that you can walk almost anywhere in Wales if you follow public footpaths and lanes. Second, it is particularly appealing to think that I'll be crossing the legendary Landsker Line, the boundary that separates the historically Welsh-speaking and English-speaking parts of west Wales. I want to delve into the history of the anglicised southern section of Pembrokeshire and learn how this part of the Welsh coast came to adopt its own particular identity.

In its simplest terms, the Landsker Line was roughly delineated by the series of castles constructed in the 200 years after the Norman invasion of Wales began in earnest in 1081. The fortifications were built both by the Norman usurpers and the Welsh princes looking to repel them.

The location of the Landsker Line fluctuated through the ages depending on military campaigns but, ultimately, it became entrenched in an area that starts around Laugharne in the east and cuts through Pembrokeshire until it reaches the seaside hamlet of Newgale to the west. Along the way, this imaginary line was fortified by castles at Wiston, Roch and Narberth among other locales, some now lost to history. Quite what this boundary really meant for the daily lives of those living in this part of west Wales is difficult to say. What we do know is that, over successive centuries, a linguistic and cultural divide grew between north and south Pembrokeshire. When the antiquarian writer George Owen visited Pembrokeshire in 1603, he referred to the land south of the Landsker Line as 'Little England Beyond Wales'.

Part of what made the Landsker Line so contentious is the way the Norman lords are believed to have reinforced their gains. According to one of the earliest written histories of Wales, *Brut y Tywysogion* (Chronicles of the Princes), the English King Henry II undermined the local population in south Pembrokeshire by importing settlers from Flanders (modern day Belgium):

> This was on account of the encroachment of the sea on their country, the whole region having been reduced to disorder, and bearing no produce, owing to the sand cast into the land by the tide of the sea.

We know from thirteenth-century records that an exceptional series of floods swept across the North Sea lowlands during that time causing great losses of land and people in the marshlands. It's quite possible that these Flemish settlers were fleeing the natural disaster that had destroyed their own lands. And it seems probable that the Norman rulers took advantage of their ancestral neighbours' misfortune to bring them into south Pembrokeshire and shore up their gains. How much this wave of climate immigration physically displaced the local Welsh-speaking populace is impossible to say. Some people must have been forced north while others would have stayed.

The net effect was the same though. Over the next few centuries, these newcomers became part of the fabric of this part of Pembrokeshire. Together with a further influx of English immigrants – mostly from the West of England – they changed the culture of these lands south of the Landsker Line. Spoken Welsh and also Welsh place names were replaced with English names (Pembrokeshire has 155 towns or villages ending in 'ton', the Old English name for a farmstead or village). This new English tongue was spoken in a dialect that, a bit like the south Gower, resembled Lyme Regis more than Llandysul.

World history has continually been shaped by migration. Wales is no exception. Inevitably, with migration comes tension but, over time, the integration of different peoples and their culture broadens and deepens our own culture. That mix is what has made modern Wales such an invigorating and dynamic place, and it excites me to be part of a nation that, for the most part, embraces the wider world. My family background mixes Welsh, English, Irish and Swedish. My children are a mix of that heritage, American and Korean. They also are fluent in Welsh – exactly the sort of cultural mix many people in Wales can identify with while still feeling Welsh.

This optimistic embrace of migration isn't shared by everyone; immigration has always been weaponised to divide and instil fear. Those tensions are only going to grow in the coming decades as climate change forces millions of people to leave their homes and homelands, as those Flemish settlers may well have done centuries before.

My Saturday morning cross-country stroll to Laugharne runs into problems almost immediately. A concrete barrier blocks access to one of the fields that the public footpath runs through, forcing me to scramble over a brambled fence and emerge with cuts and stings on my arms and legs. Cursing the abuse of this public right of way, I finally connect with the Coast Path just outside Laugharne and follow it down into the town, pausing for a few minutes to stop and appreciate the view of Carmarthen Bay that Dylan Thomas, Laugharne's favourite literary son, would have enjoyed while seated at his writing desk.

The path drops down the hill to the town's ruined castle and the mouth of the Afon Coran. The original castle, built in 1116, was reinforced (with some haste) later that century as a Norman stronghold to protect the coastal lands taken from the local Welsh population during the previous decades. It proved no match for Llywelyn the Great though who captured it in 1215. Two hundred odd years later, the last true Prince of Wales, Owain Glyndŵr, suffered a humbling defeat here, losing 700 men in an ambush and retreating after a soothsayer warned that he would be captured. It would be the start of the end of his rebellion against the English crown.

Here, at the base of the castle, it is obvious that Laugharne faces its own serious sea level challenges. The Afon Taf Estuary and the greater Carmarthen Bay have long been susceptible to high spring tidal flooding and this year is no different. Just a few hours before most of where I am now standing, including the adjacent car park and a street called The Grist, had been under a foot of water.

I spend the next hour sitting on a bench below the ruined walls of Laugharne Castle, pondering the future of this pretty but exposed estuary bay. One recent local government study identified Laugharne along with Ferryside and Llansteffan –

neighbours on the greater Tywi Estuary – as being at high risk of coastal flooding. The threat to the village will only increase in the coming years.

In front of me though is one potential lifeline – the wide sway of salt marsh and sea grass along the Taf Estuary that could provide coastal protection in the future. According to a recent study of eight Welsh estuarine environments led by academics at Swansea University, improved and restored coastal wetlands act as natural defences and can reduce flood levels from storms by up to two metres and avoid $38 million in damages per estuary.

Salt marshes, whether in estuaries or directly facing the coast, also play an important role acting as giant 'carbon sinks' – capturing and storing more carbon per hectare than many other habitats. Even when salt marsh plants die, their carbon is trapped and buried in the mud instead of decomposing and being released into the atmosphere.

The Swansea University research has particular resonance as climate change and extreme weather take an increasingly heavy toll on the world's major urban environments. Twenty-two of the largest thirty-two cities in the world – including London, New York and Tokyo – are built on low-lying land around estuaries. On a local Welsh level, the three biggest cities of Cardiff, Swansea and Newport, as well as major towns like Port Talbot and Llanelli – all sit exposed at the mouth of tidal estuaries.

A few weeks later, I find myself a few miles further west from Laugharne at Amroth beach. I'm particularly excited about this walk because my son Dylan is joining me. He has been away at university for the past two years and I don't get many opportunities to spend time with him.

I've been checking in with Dylan for the past couple of weeks (in that annoying way parents do) to make sure he could still make the trip. Now we are here, on the beach and ready to explore the coast up to Tenby. It won't be the longest walk because I know there's a storm front approaching from the west. We are going to get drenched. The question is how much ground can we cover before the rain arrives?

The wind picks up as we walk along the beachfront. We pass Bertie the Sea Bass – a tall glistening sculpture made out of waste glass and plastic by a local artist to raise awareness about marine pollution. It shows how much our appreciation of the coastal natural beauty and the importance of the environment has evolved in just over 100 years. At the turn of the last century, this expanse of coastline was a gritty industrial hub – one of the westernmost fringes of the south Wales coalfield. Today, you can still see a thick black seam of coal in the cliff face where it meets the sea.

The beachfront suffered even more abuse during the Second World War when it was used by the Allied military forces as a practice ground for the D-Day invasion. In July 1943 the beach was closed to all residents for Exercise Jantzen – a rehearsal of landing troops and supplies from the sea. Amroth was chosen because of its similarity to the Normandy beaches. Over thirteen days, military ships from Tenby and Swansea unloaded 16,230 tonnes of supplies at Amroth and it was rumoured that Prime Minister Winston Churchill came to inspect the exercise.

However, to get those supplies off the beach, the military bulldozed through Amroth's natural sea defence – a bank of pebbles at the top of the beach. In the years following Exercise Jantzen, the government had to invest in rebuilding the sea wall and laying down groynes – wooden linear sea defences that trap sand – to protect the beach. The military also left a reminder of the mass manoeuvres – an abandoned landing craft that sat stuck in the sand until, finally, the beach swallowed it whole.

We negotiate the climb up out of Amroth beach and head towards the old port town of Saundersfoot. We drop down through woodland and country lanes until we reach Wiseman's

Bridge and I stop in my tracks. Do you remember where you were when Elvis died on 16 August 1977? Well, I was here, in Wiseman's Bridge, on a summer Cub Scout camp. Seeing the beachfront again brings memories flooding back – of rows of bunkbeds, of games of football on the grass, of feeling homesick and of hearing on the radio that the King was dead.

The wind is blowing in our faces, pegging us back, as we try to progress around the headland – following the old tram tracks that once carried coal shipments from Amroth to Saundersfoot.

'We'd have been better off walking this route the other way round,' says Dylan shouting through the gale. 'We'd have done it in half the time.'

The paved tram track enters a damp tunnel that runs through the cliffs – it is pitch black apart from a trail of red floor lights to guide our way. The moisture from the tunnel roof sends a steady drip down the back of my neck.

The last few miles are a bit of a 'keep your head down and just keep moving' slog through muddy woodland trails and a long, slow and leg-draining climb over a hill near Waterwynch Bay. We are both drenched but it doesn't seem to bother Dylan one bit. He powers up the hill, headphones in his ears, while I labour my way. In that moment I can appreciate the benefit of having the legs of a twenty-year-old.

Finally, the picture postcard skyline of Tenby comes into view. Except, today, the bright orange, yellow, baby-blue and lilac façades of the buildings that look out over North Beach have been dampened dark grey – their hues muted. Thick, low grey clouds hover above the roofs. A few intrepid/daft families have established a beachhead on the sands – determined to enjoy themselves. One father and son are attempting a game of rugby. The wind makes a mockery of the oval ball each time one of them tries to kick it.

Dylan and I walk down The Croft until it joins The Norton and finally High Street. Halfway down is St Mary's Church. I wanted to bring Dylan here because this was where my paternal grandparents got married. My grandfather, a blacksmith, also forged the iron railings that surround the graveyard. He grew up in a house located on Upper Frog Street, somewhere near the De Valence Pavilion just a couple of streets away from here.

As much as I'd like to think the crafted iron railings are St Mary's principal attraction, the church has a much greater claim to fame. Inside sits a memorial to Robert Recorde, a sixteenth-century Tenby native who is considered one of the most influential people in the evolution of mathematics. He invented the equals sign (=) and introduced the Germanic plus (+) and minus (-) signs to English speakers.

The Tenby that Recorde grew up in had been a bustling and important port and fishing town for many centuries and pre-dated the anglicisation of south Pembrokeshire. Its original Welsh name is Dinbych-y-Pysgod, which translates as 'the little fortress of fish'. In the fourteenth century it developed into one of the main herring ports in south Wales and became an import centre for salt and wine arriving from France, Spain and Portugal. John Leland noted that Tenby was 'very welthe by marchaundyce' when he visited in the late 1530s. Nowadays the Tudor Merchant's House, located a few streets away from the church, offers a window into the past prosperity of the town.

Robert Recorde's genius was recognised at an early age. He left Tenby to attend Oxford University at just thirteen years old in 1525. He was awarded an Oxford licence to practise medicine in 1533 and then became an academic at Cambridge University. He went on to write groundbreaking books that would shape how arithmetic and geometry were understood, practised and taught. It was in his final book, a treatise on algebra entitled *The Whetstone of Witte*, where he explained how two parallel lines =, 'because no two things can be more equal' should become the mathematical shorthand for the sign of equality. It's hard to imagine how the modern computing power our society depends on would have developed without Recorde's sign.

Recorde also helped define how a new generation of seafarers approached navigation. Among his many talents, he was involved with the fledgling Company of Merchant Adventurers to New Lands (later known as the Muscovy Company). At the time, this group of sixteenth-century maritime entrepreneurs were trying to plot a northerly sea route to China and the Far East – the legendary Northeast Passage.

Another Recorde book, *The Castle of Knowledge*, published in 1556, was a sophisticated exploration of astronomy intended to help the Muscovy Company's navigators plot an accurate course by the location of the stars. It is a remarkable piece of work not least because it is considered the first English language book to offer a significant commentary on Nicolaus Copernicus's *De revolutionibus orbium coelestium* (*On the Revolutions of the Heavenly Spheres*) written in 1543. It was in this book that the polymath unveiled to the world his heliocentric theory – namely that the sun was the centre of the universe, not the earth, and that all planets orbited around it. Until then, academics had stood by the geocentric model of the universe credited to the Alexandrian second-century mathematician Claudius Ptolemy. It maintained that all planets and the sun revolved around the earth.

Copernicus's theory turned on its head the way humans considered the world in relation to the stars and planets. Over time it would shape our modern understanding of how the earth's proximity to the moon influences tides, currents and, as a result, weather patterns.

Recorde chose not to take sides in *The Castle of Knowledge* between heliocentricism and geocentricism. He intended to explore Copernicus's thinking in more depth at a later date. But he didn't live long enough to do so. Despite his brilliance and capabilities, Recorde died broken and penniless aged just forty-six. He had made the mistake of crossing Sir William Herbert, the Earl of Pembroke, and part of the same family that the pirate John Callis was said to run with. Herbert sued Recorde for libel and a court ordered he pay £1,000 in damages. Recorde was thrown in jail and died soon after.

After my wet and windy walk with Dylan, I decide to head off alone – exploring chunks of the southern coast from Tenby to Manorbier and Stackpole Quay to Bosherton, taking time to embrace the calm isolation of Barafundle Bay (off-season) and the spectacular cliff walks to Stackpole Head. This area is climbing country due to the challenging rock faces that rise near vertically from the sea. More than once, I'm surprised to see a head pop up above the cliff edge as a climber summits.

This section of the Wales Coast Path reaches an impasse at Castlemartin military range – a 6,400-acre area of land divided into two sections – East and West. The first allows walkers access most of the year except for when live firing drills are taking place. It includes St Govan's Chapel – the tiny stone hermitage built into the rocks where the sixth-century Welsh saint sought peace and isolation. Legend has it that when Govan was being pursued by pirates, a cleft opened up in the cliff and he was able to squeeze inside and hide. This is where he built his simple home. The East part of the range also includes Stack Rocks and The Green Bridge of Wales – natural carboniferous limestone features that have been shaped through millennia by coastal erosion.

The West section of Castlemartin, however, is off limits to civilians because of the risk of unexploded ordnance. A few times each year, though, the Pembrokeshire Coast National Park Authority runs small, organised walks across the land.

The Army first started live fire exercises in this area back in 1939 at the start of the Second World War. For centuries before, this had been agricultural land owned by the Cawdor Estate. When the military arrived the tenant farmers were forced to move at short notice. In 1948 the UK government purchased the land and, since then, only tanks and soldiers have had ready access to this western part of the coast.

The removal of most human activity has had a gradual but significant influence on the coastal landscape. On the downside, it has been pounded on a regular basis by heavy artillery shells. The upside, however, is considerable. With no pesticide use, the eradication of intensive agriculture and no human development, wildlife has flourished and, over time, both parts of the range have become a haven for choughs, marsh fritillary butterflies, grey seals and green winged orchids. A small woodland area in the West range – a rare occurrence in this denuded and exposed coastal fringe – is home to badgers, squirrels and tawny owls. Perhaps most notably, the Castlemartin range has created a protected habitat for local bee populations including the increasingly rare pale-yellow-and-grey shrill carder.

Castlemartin hasn't been rewilded. Cattle still graze the firing ranges and, each winter, sheep from the Preseli Hills are brought down to the range – reviving the once traditional *Hafod y Hendre* farming system that saw sheep move from the uplands in the summer to the lowlands in the winter. Instead it has become a vibrant – if accidental – nature sanctuary.

Milford Haven waterway looks like nothing else in all of Wales. This deepwater glacial bay – the natural extension of the Afon Cleddau – formed 12,000 years ago. Its pure size takes your breath away when you first witness it. I mean, this is Wales. We do pretty but never oversize. Our mountains are impressive but never overpowering. Our coastline has grand beaches and can be craggily atmospheric but, for the most part, it feels cosy rather than overwhelming. But this bay is something completely different. From the air it resembles a mythical creature's open mouth so far does it reach inland. From the ground, on the cliffs overlooking its southeastern flank where I am right now, I have to remind myself that this is the mouth of a river estuary.

I'm standing on a headland overlooking Thorne Island, one of a series of intimidating-looking military forts situated around the perimeter of the bay. They were built in the mid-nineteenth century because Lord Palmerston's government believed a resurgent French navy posed a direct threat to the Royal Dockyards located in Pembroke Dock a mile inland up the Cleddau.

What looks like a brown-winged Baird's sandpiper hovers in front of me before divebombing into the bay. The white-and-green-hulled Rosslare to Pembroke Dock ferry approaches the mouth of the Haven while, on the horizon, a small cluster of container ships wait for tugboats to guide them into port.

This part of Pembrokeshire has been strategically important from a military perspective for thousands of years. Looking now at the size of the bay, I can see why so many sailors, from the Vikings to Lord Nelson, valued it so highly. I take a few minutes to gaze out at the water, imagining the ships and people who have navigated this body of water in the past.

We know a Norse chieftain called Hubba wintered here in 854 with twenty-three of his longboats. The name Milford is an anglicisation of Melrfjordr – an amalgam of Old Norse *melr* (sandbank) and *fiordr* (inlet). Modern day Milford was built on the site of a small coastal village called Hubberston.

Later, in 1171, Norman king Henry II assembled a fleet of 400 warships, 500 knights and 4,000 men-at-arms in the bay before launching an invasion of Ireland across the Celtic Sea. It was also in the Haven that, in 1405, French troops landed to reinforce Owain Glyndŵr in his rebellion against Henry IV.

Glyndŵr would ultimately fail but, eighty years later, another Welsh-born royal, Henry Tudor, arrived here before marching through Wales to defeat King Richard III at the Battle of Bosworth. And it was also here that the arch anti-Royalist

Oliver Cromwell launched his own invasion of Ireland in 1649.

The Pembroke Dockyard was in operation until 1922 and in that time saw the construction and maiden voyage of five Royal Yachts and 263 other Royal Naval vessels including HMS *Erebus*, one of the two exploration ships that disappeared in Canadian Arctic waters in 1845 while trying to navigate the fabled Northwest Passage.

The Royal Dockyards are just a memory nowadays. In fact, despite Pembroke Dock's reputation for constructing ocean-going warships, perhaps the most famous vessel ever built there was a very different type of ship. In the late 1970s, long after the dockyards had fallen into decline, one of the hangars was used to build a full-sized model of Han Solo's smuggling ship, the *Millennium Falcon*, for the Star Wars sequel, *The Empire Strikes Back*. Sitting now, looking out towards the village of Dale across the bay, I can't help but think that, as exciting as it would have been to see the Viking fleet or HMS *Erebus* sail away, watching Han pilot the *Millennium Falcon* through the mouth of the Haven would have been really something.

I push on past Chapel Bay Fort towards the RNLI lifeboat station at Angle. Across the bay, two enormous jetties stretch out into the water – ready to receive the oil and natural gas supertankers that now dominate this waterway.

As Pembroke Dock's influence in the region waned, Milford's started to grow once more. It had first come to prominence (Vikings notwithstanding) in the late 1790s when the major landowner on the north side of the Haven, Sir William Hamilton, offered Quaker families from the US island of Nantucket the opportunity to establish a whaling station there.

The whaling work was a failure. Still, Milford's impressive natural harbour allowed the town to flourish as a fishing port. In the Second World War, as at Amroth beach, it acted as a staging base for US troops preparing for the D-Day marine and air assault on Nazi-occupied France. Then, in 1960, the Esso Company, the oldest foreign affiliate of John D. Rockefeller-founded Standard Oil, opened an oil refinery near the town. It was a time when the UK was decreasing its reliance on coal fuel and rapidly increasing its use of oil for heating and transportation. By the early 1980s, the Esso refinery was the second largest in the UK.

I remember, as a small child, coming to Milford Haven to visit an old family friend, Bill Fellingham. He was a ship's captain for Esso, in charge of one of the oil tankers, and he had a house that looked down over the Haven. Back then, those tankers seemed huge and thrilling as they dominated the landscape around them.

I head back inland to my final port of call today, the Old Point House. Its white stone building, on the shore of an inlet, is just the type of old coastal pub you hope to end a good walk at. The garden is packed with Easter holiday tourists enjoying a drink in the sun.

I sit among the throng, looking out onto the calm water in Angle's sheltered bay, the Valero oil refinery being the only blot on the landscape as I gaze inland back across south Pembrokeshire and the lands below the Landsker Line. This part of the Welsh coast has a particular sense of order and regimentation – a topographical hint at how it was annexed by the Normans, repopulated and then extensively cultivated over 900 years. It lacks the wildness of some other parts of west and north Wales, despite its spectacular cliffs and beaches and the recent laissez faire administration of Castlemartin firing range. That military influence, dating back to medieval times, also means this part of the coast has been one of the most marshalled and policed areas in all of Wales. Which makes sitting here at the Old Point House even more poignant. That's because this is the pub where the pirate, John Callis – a man who could give Han Solo a run for his outlaw money – would hide out when being pursued by the law. That, I think, is known as the art of hiding in plain sight.

![ornament]

7

The Sustainable Sea

Walks from Martin's Haven to Newgale and Strumble Head to Fishguard

You have to be pretty determined to start a walk from Martin's Haven. This small National Trust property sits at the far tip of the Marloes Peninsula just past Milford Haven as it rounds the corner from Skomer Island, the largest island off the Pembrokeshire coast, and opens into the wide crescent of St Brides Bay. On a very clear day, and if you have good eyesight, it's possible to see Ramsey Island, sitting some ten miles away off the coast of St Davids.

The main reason people come to Martin's Haven is to take a nature sightseeing boat trip to Skomer, which maintains one of the largest puffin populations in the southern part of Britain, or to make a pre-arranged visit to its more isolated neighbour, Skokholm.

Both provide a home for Manx shearwater (Skomer is considered the largest breeding colony in the world for the birds) while thousands of razorbills and guillemots also vie for space on the islands' cliff tops and ledges. Fulmars, gannets, shags and cormorants also frequent its shores, so you can see why birdwatchers are so keen to visit.

I'm at this outcrop for a different reason. I want to view the scenery that provided the inspiration for the Pembrokeshire Coast Path, the forerunner of the national walking trail.

The Pembrokeshire Coast Path was conceived by Ronald Lockley. He was born in Cardiff in 1903 and grew up in the new suburb of Whitchurch. He was a wanderer and a nature lover from an early age – he would go camping on his own in the wetland woodlands adjacent to the Glamorganshire canal, what we now know as Long Wood Nature Reserve. He also was fascinated with islands and the peace and serenity they could bring. He constructed a hut on a small mound of wetland – which he called Moorhen Island – in Long Wood and, as a teenager, he made a solo trip to Lundy Island where he saw Manx shearwaters for the first time.

In 1927, at the age of twenty-three, he and a friend visited Skokholm. There they met the sole tenant who quickly made clear his intention to get off the island for good. Lockley saw the opportunity to follow his dream and secured a twenty-one-year lease to live there.

Lockley had been inspired by reading Henry Thoreau's *Walden* but it's also tempting to view him as a twentieth-century manifestation of those early Celtic saints who sought solace from nature on the rocky outcrops off the Welsh coast. Unlike Cadoc and co, Lockley was married and persuaded his wife to join him. Whatever drove him to settle on Skokholm, his choice seemed so strange back in the 1920s that he was once referred to as 'the loneliest man in Britain'.

Over the next decade or so, Lockley established himself as a naturalist of international repute – becoming an expert in the breeding habits of the Manx shearwater. This was during the early years of the modern environmental movement and Lockley's observations of bird migrations, gleaned at the very edge of Wales on the Atlantic coast, made him a pioneer in his field. It was also during his time on Skokholm that Lockley first demonstrated his commitment to environmental campaigning. In 1930 he saw and documented the illegal discharge of oil by a passing tanker. This would lead to the first successful pollution prosecution under the 1922 Oil in Navigable Waters Act.

Lockley was forced to leave Skokholm in 1940 when the Army took over the island during the Second World War, but he continued to live in Pembrokeshire. In the 1950s he took a leading role in setting up the Pembrokeshire Coast National Park. It was during this time that he hit on the idea of creating a walking path round its 186-mile coastline. It would take seventeen years before the path was completed and involved years of intense negotiation with private landowners to gain access to their land.

Not surprisingly, Lockley was vehemently against the decision to develop Milford Haven as a major oil terminal hub. Having seen his opposition ignored and rebuffed, Lockley decided to emigrate to New Zealand in 1970, convinced that the threat to the natural world from oil and other industry wasn't being taken seriously. He spent the remainder of his days travelling around Polynesia. Even if you've never heard of R. M. Lockley, you'll likely know one part of his legacy as a prolific author. In 1964 he wrote a book called *The Private Life of the Rabbit*. It was the inspiration for Richard Adams's novel, *Watership Down*.

The plan is to follow in Lockley's footsteps and walk a long stretch of the Coast Path around St Brides Bay. I've convinced another friend, Geoff Jones (not to be confused with the other Jeff Jones), to join me.

The sky is refreshingly blue as the great expanse of St Brides Bay comes into view. Across the water, it is just possible to make out the beach at Newgale. It lies fourteen miles away and looks a very long walk now I can see the entire route laid out before us.

'I think we can do this in six hours or less,' I say confidently – partly for my own encouragement.

Geoff doesn't say anything. Instead, he nods in a serious sort of way that could be interpreted either as a full agreement of my estimate or polite prevarication on account of never having walked

Seascape

with me before (but having heard stories of those who have).

We maintain a steady pace over the firm, single-track dirt path that cuts through the gorse on the top of the cliff. Below us, a lone fishing boat ploughs a slow but steady route across the gentle sea.

This bay is renowned for pods of dolphins and for the seals that hug the rugged coves.

'There's a seal below us,' I say, pointing to a bobbing black dot in the water.

'Are you sure that's not a buoy?' Geoff replies, adding, 'You'd expect a seal to, well, move more, no?'

He is, as with all interactions, polite to a fault – even when shooting holes through my wildlife expertise.

Geoff is a strong walker with a steady, deliberate gait. He is a tall man and his steps eat up the ground with an efficiency of effort. More important, he is easy company to walk and chat with, so the morning passes quickly as we track along the Coast Path.

Out in the bay, a red-hulled fuel tanker with a white-painted deck has peeled off from the queue further out to sea and approaches the turn around Skomer Island. The presence of maritime heavy industry is a normal sight here given the proximity of Milford Haven. It is also a reminder of the risks we continue to take with our coastline and our planet to feed our fossil fuel habit – exactly what Lockley was concerned about. Most of the time, the pollution that comes from our oil use goes into the atmosphere, making it impossible for most of us to quantify. Occasionally, though, we get to see the direct effects and when that happens, the impact is dirty, visceral and deadly.

In February 1996, an oil tanker named the *Sea Empress* ran aground on rocks off St Ann's Head at the mouth of the Milford Haven Waterway, just around the bay from here. Over the next week the stricken tanker spilled 72,000 tons of crude oil into the sea. It was the third largest spill in the UK and it took place within Pembrokeshire National Park, threatening wildlife and a growing tourism industry.

The fallout was immediate. Over 125 miles of coastline were polluted and more than 7,000 birds perished or were coated in crude oil. One of the most impacted sea creatures was the cushion starfish, mainly found in the rockpools of West Angle Bay. There, the population dropped from 150 to thirteen after the spill.

It could have been much worse. Most migratory birds had not yet returned to Skokholm and Skomer. Because the prevailing wind was initially blowing from the north, just a small amount of the total oil spilled reached the coast.

Today Welsh maritime wildlife faces a far greater challenge than one oil tanker. Warming ocean and air temperatures are changing the balance of nature and the types of species that can thrive, or just survive, along the Welsh coast.

In 2023, ocean temperatures around some parts of the United Kingdom were five degrees warmer than have traditionally been recorded. This is putting marine life under incredible stress. Rising sea temperatures are disrupting the food chain of sea birds like the Manx shearwater and puffin. Both prey on sand eels who in turn eat zooplankton. The latter struggle to survive in warmer waters.

The warming seas are also attracting new visitors. Already this year, a giant Portuguese man o' war jellyfish has washed up on a north Wales beach and a young sperm whale beached at Porth Neigwl – only the second of its kind to have been discovered on Welsh shores. As new species arrive, they will alter the balance of wildlife that has existed for thousands of years.

We push on, following the boundary wall around St Brides Castle – less a castle really than some sort of wedding-cake-styled stately home (a mix of Scots Baronial, Tudor, Edwardian and Gothic influences apparently) that was built in the early nineteenth century.

At the craggy outcrop named Huntsman's Leap we begin a descent into St Brides Haven, a red shale rocky bay in the shadow of the comedy castle. The sea is a deep green away from the shore. Closer to the rocks, refracted streaks of light green and blue break through the darkness.

We arrive at one of most popular sections of the Wales Coast Path. This stretch around the bay from St Brides Haven takes in the villages of Little Haven, Broad Haven, Newgale, Solva and finally St Davids – all destinations to which tourists flock in their thousands during the spring and summer high season. A lone black crow perched on a white rock observes our entrance into Little Haven, almost as if it is taking a head count of the tourists wandering through town.

Meeting such a large number of other walkers gets us chatting about the joy of finding a deserted path and about how far you need to venture to shake off the maddening crowd. Both of us agree that it takes walking at least an hour away from the popular beaches before you are guaranteed a slice of serenity.

'A really decent climb also helps weed out the casual strollers,' says Geoff as we look at a particularly steep set of wooden steps, installed to help overcome coastal erosion, that will take us on to Druidston and then Newgale.

Parts of Pembrokeshire, notably Tenby, Little Haven and Broad Haven, have been seaside destinations since Victorian times and the early days of salt-water bathing holidays. In recent decades however, and especially since the COVID lockdown of 2020, this part of west Wales has almost become a victim of its own natural beauty.

In 2021, seven million visitors came to Pembrokeshire – pumping some £590 million into its economy and providing employment to 12,000 people. While the revenue is very welcome, the volume of visitors has created new problems for a part of the world used to living a quiet, mostly rural, life. For large chunks of the summer, Pembrokeshire is overcrowded with traffic clogging the narrow roads and lanes. Holiday lets and camping pitches sell out months in advance. The seasonal influx of tourists puts additional pressures on local services and amenities – so much so that the Welsh government has proposed charging a tourism tax for holidaymakers visiting, as many other countries and cities around the world already do.

More than six hours after we began the day's walk, the long expanse of Newgale Sands comes into view. A light sea mist is just starting to form out on the horizon. At the top end of the beach, we can see the Duke of Edinburgh pub enticing us to come in. It's become famous in recent years for being on the front line of climate change as ever intensifying spring tides overpower the pebble sea wall and pour into its downstairs bar. There's even a notch on the door of the pub showing how high the water reaches when it invades. The Duke of Edinburgh will have to wait for another trip, however. Today, the cathedral city of St Davids beckons.

St Davids is a place I have visited many times but it always gives me a bit of a thrill to walk down The Pebbles from the main square (more a triangle to be exact) and be greeted by the sight of the majestic cathedral. It sits down in a valley next to the Afon Alun, surrounded by undulating fields leading off to the coast with the hills of Carn Llidi, Perfedd and Penberry in the distance.

Today the low sun reflects off the cathedral's deep hued sandstone façade. A few tourists take photos of the building while, in the churchyard, a group of local teenagers mull around, doing their best to stifle the boredom of an afternoon in the UK's smallest and sleepiest city.

The cathedral sits on the site where Dewi Sant, patron saint of Wales, first established a monastery back in the sixth century. The city is named after Dewi and has been a destination for visitors ever since the Middle Ages when Christian pilgrims came to honour the saint. If they visited twice, they saved themselves a trip to Rome (though Baruc's tomb at Barry Island still seems the budget option).

Seascape

Dewi Sant's mother was also an important religious figure. According to Rhygyfarch, the eleventh-century Bishop of St Davids who wrote *The Life of St David*, Non was the daughter of a powerful warrior who was raped by the King of Ceredigion. She is said to have given birth on the cliff top near here during a fierce storm. A bolt of lightning shot down from heaven and split the rock on which she lay at the very moment Dewi entered the world. That stone now lies under the altar of St Non's Chapel.

I've got lost in my head wandering around St Davids and have also lost track of time. I'm due to meet up with Geoff in the Farmers Arms which, luckily, is only a few minutes' walk away. When I arrive, he is sitting chatting with some friends (Geoff has friends in every pub he sets foot in and if he doesn't, he'll make some by the time he leaves) so I get a drink and join them.

His companions are a father and daughter, Owen and Megan. Owen is co-founder of Câr-y-Môr, a community-owned commercial seaweed and shellfish business located just a mile or so out of St Davids. Megan works with Owen and shares his dream of creating a sustainable and regenerative ocean farming business.

Over a couple of beers, Owen recounts how he worked in commercial fish farming for many years but became disillusioned with the way the industry operates. By applying his expertise to seaweed, he is convinced he can create a business that can help mitigate climate change (seaweed can be used as a substitute for oil-derived plastic), employ local people and improve people's diets.

Câr-y-Môr has clearly caught the imagination of local people. It was created back in 2017 as a Community Benefit Society (CBS) – a business owned and democratically run by its members. Today 120 of them dedicate two days of their time a month to building the business.

The aqua farm uses a sustainable polyculture system where kelp and shellfish are grown together without the need for fertiliser, feed or fresh water. It also is helping preserve and regenerate two of Wales's historic sea foods.

The first is oysters – once so plentiful along Welsh shores that they were considered simple food, not the delicacy we think of today. The Romans imported Welsh oysters – both Caldey Island, off the coast of Tenby, and Milford Haven, with its huge coastal ria, were rich in oyster beds. On Gower, Oystermouth earned its name because of the abundant supply. In the nineteenth century, oysters were a regular part of the Welsh diet but overfishing to serve the rapidly growing population of industrial south Wales destroyed their habitat.

An oyster revival is also taking place in other parts of Wales. Back at Angle Bay, in the shallow waters outside The Old Point House, the owners have undertaken a sustainable cultivation project to boost the *Ostrea edulis* oyster population. Similar initiatives are taking place near Oystermouth and in north Wales on the Conwy Estuary.

The second, of course, is laver, a Welsh staple for thousands of years. Laver is a dark reddy-brown and green type of seaweed that is native to the western shores of the British Isles. It is rich in vitamin B12, iron and iodine. When cooked slowly – normally for ten hours or more – it effuses a deep, salty umami flavour that makes you either love it or hate it.

Carwyn Graves, in his book *Welsh Food Stories*, quotes the English historian William Camden's recollection of laver harvesting in this area back in 1607. Traditionally, the laver was fashioned into cakes, fried with oatmeal then eaten as an accompaniment with bacon and cockles. Or it was used as a spread on bread. Camden wrote:

> Near St Davids, especially at Eglwys Abernon, and in many other places along the Pembrokeshire Coast, the peasantry gather in the Spring time a kind of Alga or seaweed, where they made a sort of food called lhavan or llawvan, in English, black butter.

Today, laver is still eaten across Wales but it's no longer the peasant food of yore. And it's no longer produced all along the Welsh coast. Instead, it has been reinvented as an artisanal delicacy – the Welsh caviar as it's sometimes called – and is served in high-end 'beach to fork' restaurants. I like to think of it as locally sourced Marmite.

Other innovative food thinkers and chefs are re-introducing Welsh diners to the taste of the sea through foraging. At Narberth restaurant Annwn, chef Matt Powell serves a range of different seaweeds, including laver and rack seaweed, sea grasses and shoreline plants that are collected fresh from the local beaches and then marinated, soused or served in soups or as an accompaniment to other local dishes. Further north on the Dyfi Estuary, chef Gareth Ward is using local ingredients fused with Japanese flavours at Ynyshir while, here in St Davids, another restaurant, The Really Wild Emporium, also is showcasing foraged, local Welsh sea plants and vegetables.

By highlighting the appeal of local coastal ingredients, and by using them sparingly and sustainably, this new generation of chefs and food producers are helping reconnect people's palettes with tastes and foods that are core to Welsh heritage (even if they are being creatively reimagined). Admittedly, these are all high-end restaurants but, as with all food trends, creativity has a habit of filtering down into mainstream tastes and cooking. Hopefully this new coastal food revival can build a better appreciation for the sea and how it has sustained society in the past.

If Câr-y-Môr and Annwn can give laver a fresh lease of life, it would seem only fitting given how laver did the same for Japanese nori seaweed more than half a century ago. Nori is particularly important to Japanese cuisine and culture because it is the seaweed used to wrap sushi. Seaweed farmers had been cultivating nori since the 1600s but, in 1948, a series of typhoons, along with increased coastal pollution, caused the decimation of the industry.

Despite centuries of using nori in cooking, little was understood about the life cycle of seaweed and no one in Japan knew how to grow replacement nori plants. Nori, however, has a seaweed cousin, *Porphyra umbilicalis* (Welsh laver) and it just happened to be the research subject of a scientist at Manchester University named Kathleen Drew-Baker.

A year after the typhoons hit Japan, Drew-Baker published a short paper in the journal *Nature* in which she made the case that another microscopic form of algae, conchocelis, was actually baby *Porphyra* and not a separate species as was thought. Inspired by her research into laver, Japanese scientists were able to apply the findings to its cousin and rehabilitate the nori industry. Even today, Drew-Baker is known as the 'Mother of the Sea' in Japan.

Andy has a few days' free time and is particularly keen on walking a stretch of the Coast Path that leads into Abergwaun (Fishguard) because of his family connections to the town. So that's where we are headed today.

We arrive at Strumble Head in bright sunshine – even if there is a thirty-mile-an-hour wind blowing off the Irish Sea. In front of us stands the iconic whitewashed round stone lighthouse. It was constructed in 1908 to warn mariners about this dangerous section of the Pembrokeshire coast – more than sixty vessels had been lost along it in the nineteenth century alone.

Sometimes walking around the Welsh coast can feel a little repetitive – relentlessly beautiful if you can imagine that. Parts of Gower and Pembrokeshire are so picturesque that, and I am almost embarrassed to admit it, I get a bit bored. Here though, on the north tip of Pembrokeshire, the fluctuating blue, green and grey shades of the sea mesmerise.

We move swiftly this particular morning, our spirits buoyed by the fine weather and the dramatic views that the high cliffs provide of the sixty-two-mile-long, croissant-shaped Cardigan Bay.

At Porthsychan Bay we stop to watch a pair of seals playing in the sea. They take turns to swim close to shore then pop their heads out of the water so they can have a long, good look at us. The seals then disappear under the water before returning to the middle of the bay. It feels like their game of peekaboo – they are toying with us and it is a privilege to be part of it.

For one moment I worry that maybe we are intruding on a beach where baby pups might be sheltering. There have been stories recently about tourists getting too close to seal families on the local beaches and scaring them. There was even one reported instance of tourists throwing stones at the young pups. Andy and I scan the beach around us. It's just us and the two seals in the water. We pose no threat.

Sitting in that sheltered bay, I have time to reflect on just how much the sea calms me. When I started this walking journey back in Chepstow, it felt like hard work. I was forcing myself out into nature to sup on its tonic. Now the walks come easy and I get a sort of childish excitement at the prospect of escaping regular life and heading out to the coast. That said, I'm surprised that it's taken this long for me to relax back into the rhythm of walking and being at one with the outdoors. Sometimes it's easy to fall out of a healthy routine even when you know it is essential for your well-being. Then you find yourself sitting on a beach in the middle of nowhere, with no other distractions (and no phone signal). And you realise this is where you belong.

We climb out of the seal bay and walk until we reach a stone monument at Carreg Wastad Point, overlooking Aber Felin Bay. It commemorates surely one of the strangest episodes in British military history – the last time these islands were invaded by a foreign power.

On 22 February 1797, a French navy ship, packed with 600 soldiers and 800 press-ganged prison inmates, landed here just south of Fishguard. The ship, captained by an Irish-American veteran of the War of Independence called William Tate, had intended to dock in Bristol. There, for some reason best known to them, the French plotters thought they could persuade local sailors to mutiny against the British crown. But bad weather prevented them from sailing up the Bristol Channel so they headed further west towards Fishguard.

If this plan already sounds a bit hare-brained, it became even more so once the soldiers and criminals disembarked and set up camp near the village of Llanwnda about half a mile inland from the coast. The convict soldiers were undisciplined and hungry so, when Tate dispatched some of them to collect supplies and spread the good news about the invasion, things quickly and disastrously went wrong.

First, some of them assaulted local villagers and ransacked their houses. Next, one group of men got food poisoning – the suggestion is that they cooked stolen geese in rancid butter. Finally, to round off a bad night, some of the plunderers stumbled upon barrels of wine that the locals had recently 'liberated' from a Portuguese shipwreck. The French proceeded to drink themselves into a stupor. By the next morning a large part of Tate's army was in no fit state to fight anything apart from a severe hangover.

By now the French ship had left harbour, leaving the invaders marooned. Taking advantage of the general ineptitude, a group of local women, led by a cobbler named Jemima Nicholas, donned traditional Welsh red-and-black rural dress and, brandishing pitchforks, rounded up twelve drunk French invaders who had mistaken them for British soldiers.

Tate soon surrendered and the invasion was over before it really began.

We reach the outskirts of Goodwick and follow the footpath as it snakes down the cliff to a long breakwater and the ferry terminal. On the other side of the bay is the town of Fishguard.

Walking this section of Pembrokeshire has been a real education in the tensions that exist in modern coastal tourist communities. I've been able to explore the origins of the Wales Coast Path and the environmental commitment that helped bring it to life. I've also seen how developing tourism along this beautiful piece of coastline in a way that is sustainable for nature and the local population is a delicate balance. At times, you feel that Pembrokeshire's appeal might also be its downfall. But I've also seen how a new breed of creative thinkers are helping teach locals and visitors alike how to reconnect with nature. Just when you think that Wales might have forgotten how its character was defined by the sea, it's heartening to see how those connections are being rekindled.

There's just one more place to see on this walk – Popty Café on Fishguard's High Street, just across from the Royal Oak hotel where Captain Tate surrendered back in 1797.

Two generations ago this was where Andy's grandparents ran the corner shop. He stayed here on family holidays when he was young but hasn't been back to Fishguard in a few years so is keen to see what has become of the place.

We pop in for a coffee and he introduces himself to the owner. 'Oh, Annie Francis. Everyone used to say how much they loved coming into her shop,' she says, remembering Andy's gran.

'My gran and my mum were working here when they filmed *Moby Dick* in Fishguard,' Andy tells me as we sip our coffees. 'They had a notebook of autographs from the cast.'

That was back in 1954 when the director John Huston chose Fishguard's old harbour as a location for the film version of Herman Melville's epic novel. The famed director even commissioned the creation of a model seventy-five-foot, twelve-ton white whale that was towed out into the harbour – until it broke its moorings and floated out to sea.

The stars of the movie, including Gregory Peck and Orson Welles, stayed in Fishguard during the filming. 'My mum even got to meet Gregory Peck while he was here,' says Andy, adding: 'He used to pop into the shop for a packet of cigarettes.'

'Did she ever tell you about it?' I ask

'Every Sunday over lunch!' he says with a laugh.

8

Stories of the Sea

Walks from Cwmtydu to Aberaeron and Aberystwyth to Borth

Getting to Cwmtydu by road involves a winding, single lane descent that tests the patience and switchback skills of motorists but is well worth the effort for the reward at the end – a dramatic V-shaped bay that, a few hundred years ago, was a centre for lime making. The lime was exported up and down the coast to help farmers in south and north Wales fertilise acidic land. Cwmtydu is also a favourite bay for grey seals who nurse their calves on the beach at low tide from August through to December.

The ruin of the old, rounded stone lime kiln sits at the top of the pebble beach where patches of wet sand reveal themselves with every retreating wave. Jeff and Tim are with me today. The plan is to walk a stretch of the Coast Path taking in two significant seaside towns – New Quay and Aberaeron.

Half a mile or so west of Cwmtydu, we pass the ruins of Castell Bach, an Iron Age promontory fort built on an islet just a few metres from shore. Over thousands of years, the wind has eroded the structure into a conical dome. From here on the cliff, it resembles a giant spinning top.

Castell Bach is just one of many promontory forts that archaeologists have unearthed in recent times – a pattern of strongholds and urban dwellings that point to the importance of the sea routes thousands of years ago. It was most likely an important trading post with other communities up and down the Welsh coast as well as other Atlantic-facing nations.

One leading authority on Iron Age hill forts, Toby Driver, suggests that Castell Bach may also have been considered a 'sacred, liminal place' because its location was so close to the sea and a possible entrance to the Annwn (otherworld). In what is now Scotland, Hebridean people believed that seals were the emissaries of the spirit world, while other Scottish and Irish mythologies celebrate the selkies – creatures that were half human and half seal. Maybe it is more than coincidence, then, that Cwmtydu is a favourite spot for Ceredigion's seals.

Whatever Castell Bach's mystical qualities, its historical pedigree is in no doubt. Which is why it is one of a number of hill forts currently being surveyed and studied to assess how quickly they are succumbing to coastal erosion and climate change. As with the ancient stories still hidden in the mud of the Gwent Levels, archaeologists and historians are trying to preserve and learn as much about places like Castell Bach as possible before the sea swallows their secrets.

The bays come less frequently now and the path follows a series of impressive bluffs that drop vertically down into the ocean. In the far distance, we can start to make out the contours of the Llŷn Peninsula in north Wales. Somewhere, beyond the horizon, lies the coast of Ireland.

Perhaps it's the vastness of the sea on the west coast of Wales – and the mystery of what lies beyond the horizon – that has inspired so many tales of giants, gods, sorcerers and sunken kingdoms.

A few miles south of where we are walking lies the village of Llangrannog. At the far end of its inviting sandy beach sits a wonky jagged outcrop of rock perhaps twenty metres high. It looks like a naturally formed Tower of Pisa. It's called Carreg Bica and legend has it that once there lived a giant named Bica who was suffering from a particularly bad toothache – oral hygiene probably not being a pressing concern for giants back then. Bica spat out the tooth and it fell into the sand forming the rock.

There are plenty more stories of giants as you move up the Welsh coastline but the most famous one has to be Bendigeidfran or Brân the Blessed. He is one of the main characters from the *Mabinogi*, the compendium of stories, legends and romance tales that were told by the bards at court during the Dark Ages and written down in the fourteenth century in two books, the *Red Book of Hergest* and the *White Book of Rhydderch*. In the nineteenth century these texts were translated into English and published as what we know today as *The Mabinogion*.

In the Second Branch of the *Mabinogi*, we find Bendigeidfran, the son of Llŷr – god of the sea and King of Britain – sitting by a beach near his court at Harlech (which means beautiful rock) in Gwynedd. He spots thirteen ships approaching from Ireland. They belong to the Irish king Matholwch, who has come to woo Branwen, Bendigeidfran's sister.

The two hit it off and get married. Branwen returns to Ireland with Matholwch where they have a son, Gwern. Soon, though, Matholwch begins to beat and mistreat Branwen – punishment because her half-brother Efnysien had mutilated the Irish king's horses in anger at not being consulted about his sister's wedding. Keen to escape, Branwen sends a starling back to Wales with a message asking for help. Bendigeidfran assembles a mighty war fleet and sets sail for Ireland. Being a giant, he leads his warriors by walking ahead of their ships through the sea.

The mission is what military commanders today might call a qualified success. Bendigeidfran succeeds in rescuing Branwen

Seascape

but she dies of a broken heart after Gwern is brutally murdered by Efnysien (who clearly has anger issues). Bendigeidfran, meanwhile, is mortally wounded and, to make matters worse, only seven of his knights survive. He instructs them to cut off his head and carry it back to Wales where they live with it for eighty-seven years, first on the Llŷn Peninsula and then on the Isle of Gwalia (thought to be modern day Grassholm). For at least the first seven years, Brân's severed head remains able to chat with the knights to keep them company.

The *Mabinogi* also tells us of Dylan, known as the son of the sea. He was the son of Math and Arianrhod who magically gave birth to him after stepping over the wand of Gwydion, a mischievous and malevolent sorcerer. The text describes how, upon being baptised, Dylan made for the sea and once immersed, 'took on the sea's nature and swam as best as the best fish in the sea.' We later learn that Dylan travels up the coast with Gwydion in a ship conjured out of seaweed. Yet, despite his superhuman abilities, Dylan is accidently killed by his uncle, another deity called Gofannon fab Dôn. It's said that the roar of the tide at the mouth of the Afon Conwy in north Wales is Dylan's death groan.

After two hours of walking, twin bays come into view. The seaside town of New Quay sits nestled below the cliffs – a patchwork of brightly coloured cottages ascending from the water's edge.

It is the start of the summer holidays and, down in the town, the narrow streets are packed with families looking for amusement on a day when the winds offer little hope of beach fun. At the quayside jetty, families queue up for dolphin watching trips out in Cardigan Bay. Others seek dry land succour at the local chip shops and ice cream stalls. The whole town bustles with the pent-up frustration of a British seaside town on a soggy July day.

There is another legend associated with Dylan and it takes place here in New Quay. Frankly, he doesn't come out of it looking that great.

Here's how it goes. Branwen, Gwenllian and Llio were the

beautiful daughters of a wealthy local man. Every day they would walk along the beach. As they did, the waves would whisper: 'Branwen, Gwenllian, Llio – the prettiest girls in the world.' The sea carried news of their beauty down to the depths where Dylan lived.

Intrigued, Dylan swam to New Quay to see the girls and, when he raised his head above the water, he was amazed. That night, in his castle on the seabed, he resolved to bring the three sisters to live with him.

The following evening a powerful storm whipped up the waves along the beach and washed over the fishermen's boats in New Quay harbour. Branwen thought she heard the sound of someone calling outside her house but when she went to the door to investigate it slammed shut behind her and she vanished. Gwenllian went to check on her sister and the same thing happened. Llio, worried about her sisters, opened the front door and immediately felt the sea washing over her feet and foam on her cheeks. Then came a hand on her shoulder – it was Dylan.

He took the girls down to the depths to live with him but soon realised they were not happy – they longed to be back with their father in New Quay. Dylan was sad because he knew they were unhappy but he also knew they could never return in human form after the spell he had cast on them. So he came up with a plan to make the girls both part of the sea and part of the land at the same time. He transformed them into beautiful white gulls. That way, they could visit their father but still return to him.

Each day, the girls' father would walk on New Quay beach looking for his daughters and, each day, three gulls would keep him company. Finally, he realised that the three birds were his daughters and it broke his heart.

Some say that is why the cry of a gull is so sad – because it is longing for 'Branwen, Gwenllian, Llio – the prettiest girls in the world.'

Seascape

We pause in a small park above New Quay harbour. A statue of a mermaid blows a kiss out to sea. It was created by an artist, David Appleyard, to mark the official midpoint on the Wales Coast Path. Appleyard wrote when explaining his vision:

> Parallels have been drawn between a seafarer's heightened awareness and knowledge of the wind and weather and that of a person traveling the Coast Path. Not only in their preparations for a voyage, walk or run but also in their awareness and appreciation of the ever-changing land and seascapes.

'Clearly he's never gone walking with you,' says Tim.

A second monument celebrates the life of Dylan Thomas who lived in New Quay for a few years during the 1940s.

Thomas's father named him Dylan after the mythical son of the sea. Back then, in 1914, there were almost no other children with that name. Today Dylan is one of the most popular names not just in Wales but also in America – though a certain folk singer changing his name from Zimmerman to Dylan in 1962 in homage to the poet probably had a lot to do with its growth in popularity.

Most people consider Laugharne to be Dylan Thomas's spiritual home – it is there at the Boathouse where his writing shed is preserved. But literary detectives argue that many of the eccentric and enchanting characters in his best-known play *Under Milk Wood* were inspired by the playwright's experiences and encounters while living here in New Quay.

The fictional fishing village of Llareggub ('Bugger all' spelt backwards) could have been based on any number of declining coastal backwaters when Thomas wrote it over seventy years ago. It is, in many ways, an everyman place for a coastal Wales clinging on by its fingertips to a fast-fading way of life – one that was being undermined and usurped by the growth and allure of modern city living. Today, *Under Milk Wood* reads as both a satire of snooping and sniping Welsh village life, and yet also an homage to a tight-knit community fuelled by sadness, humour and, despite the murder fantasies, a sense of caring.

Our walking route heads up the main street before cutting through the garden of the Black Lion pub, a favourite haunt of Thomas. It leads onto Brongwyn Lane – the route he would have followed home on his nights back from the pub (though perhaps not in a straight line).

The tide is low at Traethgwyn beach so we stroll across the soft sand towards the end of the bay. Above us, much of the scrubland at the top of the beach has collapsed. Widespread erosion is eating away at the shallow cliffs, exposing tree roots and layers of sandy rock and hard mud. A wooden walkway lies in tatters where the ground has disappeared below it.

A cluster of small boulders is scattered in deep, soft sand at the top of the beach. It looks like a party of Druids have been making standing stones. Or perhaps it might be a mermaid's playground?

According to *Folk-lore of West and Mid-Wales*, written by travel writer Jonathan Ceredig Davies and published in 1911, it was here that a mermaid often sunned herself on the rocks. One day, while swimming out in the bay, she got caught in some fishermen's nets. She begged the men to cut her free and they did so. In return, the mermaid warned the fishermen of a storm that was brewing further out to sea and urged them to head back to land. They paid heed to her warning and returned safe before the storm hit. Other local fishermen weren't so lucky. One boat and its crew were sunk that day.

The book also recounts a similar tale published in *Y Brython*, a Welsh-language newspaper. It tells of a no-good fisherman named Pergrin who encountered a young mermaid 'doing her hair' on a rock near Aberteifi (Cardigan), about twelve miles south of here.

Pergrin took the mermaid prisoner, but, luckily for her, she was fluent in Welsh and able to negotiate her release by promising to warn her captor if he was in danger.

'Pergrin, if thou wilt let me go, I will give three shouts in the time of thy greatest need,' she told him.

He released her and she disappeared into the depths of Cardigan Bay. Weeks passed until one calm day, Pergrin was fishing when the mermaid appeared out of the water crying: 'Pergrin! Pergrin! Pergrin! Take up thy nets! Take up thy nets! Take up thy nets!'

He did as she instructed and, by the time he had returned to shore, a violent storm engulfed the bay. As *Y Brython* wrote: 'Twice nine others had gone out with them, but they were all drowned, without having the chance of obeying the warning of the water lady.'

Mermaid stories abound the world over wherever coastal and fishing towns exist. Sometimes mermaids are destructive forces – using trickery and so-called feminine guile to lure sailors onto the rocks or into deadly storms. These two Welsh *morforwyn*, however, seem far more benevolent, protecting the men from the dangers of the sea even when they, clearly, have been treated badly.

We continue until we reach the mouth of a small stream. A sign says that the quality of the bathing water is being monitored by the Welsh environmental agency, Natural Resources Wales. There is a chance that run-off water from this stream and others running into the bay may 'give rise to water quality issues'.

This is a euphemism for sewage pollution – an issue that has blighted the British coast in recent years. In 2023 alone, untreated sewage was discharged into the sea more than 399,000 times by water companies. Raw sewage was dumped 579 times and for 6,757 hours on to Welsh 'Blue Flag' beaches, which are supposed to be the cleanest and safest in the UK.

The water companies often blame the weather for having to pump sewage into the sea. They say that the volume of rain that has fallen on the UK in recent years has overwhelmed the drain system leaving them with no choice but to discharge into the ocean. But environmental watchdog groups have documented plenty of times when these utilities have dumped waste into our rivers and seas on dry days. It doesn't take a cynic to suspect that these companies are prioritising profits over the well-being of people and the environment. And it also doesn't bode well for a future where climate change will bring even more rain to Wales.

At present, it is a scandal that has shocked the public and made hundreds of swimmers and surfers ill. I, like many other people, have stayed out of the water because of the risks posed by this dumping of human waste.

Surely the mermaid of these rocks would shed a tear if she knew how her beach is being treated.

We head inland, past the old Plas Llanina mansion, one of the places where Dylan Thomas stayed and wrote during his time in New Quay. Next to it sits St Ina's Church – dedicated to the sixth-century King of Wessex who, it is said, was shipwrecked at nearby Cei Bach beach when his ship was caught in a violent storm. Ina was so well cared for by the locals that he vowed to return. When he did, he built this church, though the original building has long since been lost to the sea.

The Coast Path rises through a wooded grove towards the town of Aberaeron. The track is transformed from dirt and mud into a blanket of deep green grass as it emerges onto the open headland. The sun feels warm on our faces and our mood is upbeat.

We drop down into Gilfach yr Halen (Salt Creek). Some believe this tiny bay got its name because it was a haven for salt smuggling back in the seventeenth century. Here we make an unexpected and most welcome discovery – an appealing, white-walled building with benches to sit outside that looks very much like a pub. Albeit one with no name. We all agree that having walked past the fabled Black Lion without stopping, we shouldn't do so again. We order a drink and rest our feet. An old, weather-beaten farmer sits at the next bench, his sheepdog patiently waiting on his red all-terrain vehicle a few feet away. The farmer's name is Mal. He introduces his dog as Mei.

We ask Mal if Mei is a good sheepdog? Mal snorts. 'Sheep? He's scared stiff of them… Cattle, however. He'll happily herd them all day.'

We finish our drinks and give Mei a gentle pat as we walk away. It's hard to resist the charms of an eccentric cowdog.

The town of Aberystwyth is famous (in my mind) for a few reasons. The first is its vibrant university where the students make up for the small-town nightlife with an enthusiasm to party that rivals far larger university cities. The second is its beachfront promenade. Back in the nineteenth century, Aber (as all the locals refer to the town) was called the Biarritz of Wales because of its elegant seafront. The final reason is that Aberystwyth is home to the National Library of Wales – custodian of some of the nation's most important literary history and manuscripts.

Among the National Library's treasures is the *White Book of Rhydderch* and the *Black Book of Carmarthen* – the former written by the monks of Ystrad Fflur back in the thirteenth century and the latter featuring the earliest mention of Cantre'r Gwaelod (the Sunken Hundred), a kingdom lost to the sea that once lay off the coast of Cardigan Bay. The library also houses *Brut y Tywysogion*, where I'd learned about the Flemish invasion of Pembrokeshire and the *Book of Taliesin*, which includes poetry said to have been written by the sixth-century bard of the same name, as well as *The Elegy to Dylan – Son of the Wave*.

I did have plans to visit the library but, as I arrive in town (Jeff and Tim having headed back to south Wales), I find myself getting sucked into the summer holiday spirit. The old Victorian pier – now transformed into one giant pub – is packed. Swimmers and paddleboarders are enjoying the becalmed, waveless tranquillity of the bay. Back on dry land, the ice cream shops are doing a roaring trade.

Should I head to the library and immerse myself in Welsh legend, history and mythology? Or should I just soak up the atmosphere of Aber on the warmest day of the year? The books of antiquity will have to wait for another trip.

The next morning, refreshed by the salt air and a good night's sleep, I head north to the town of Borth. It is one of the shorter stretches of the Wales Coast Path – just five miles – but it makes up for its brevity with a lung-sucking series of climbs and descents along the steep cliffs.

This early morning promises to be another warm summer day. For the moment, though, there is a slight chill in the misty air. A lone swimmer ventures out into the bay – her heavily tattooed arms the only colour in the monochrome grey sea. Further offshore, the local lifeboat crew is running drills, practising a speedboat rescue of a swimmer in distress. A lone herring gull hovers above – perhaps looking for any prey churned up by the lifeboat's wake.

I follow the tarmacked path that zigzags its way up Constitution Hill, traversing the Victorian era Electric Cliff Railway that provides tourists a more direct (if not always quicker) route to the top. From here, I can see the grand, white classical façade of the National Library of Wales. Sitting high on a hill above the town, I can see why it's often been referred to as the Welsh Parthenon.

I pass through Clarach Bay caravan park, complete with a tattered old amusement park and mini golf course, before heading due north towards Borth along the cliff path. Below me, a man sits on the prow of his white boat, contemplating a day's fishing. Further out in Cardigan Bay, two young guys on jet skis churn up the grey-green sea with white wake foam. Rabbits dart from burrow to burrow along the cliff path and a pair of stonechats, brown feathered and yellow breasted, chit and chut as they feast on insects in the gorse.

I reach an old white house overlooking a wide and shallow bay known as Wallog. It seems odd there's a house here all by itself but the bay is even stranger. The water appears to be two different colours – one dark and one much lighter. Only later do I realise I'm looking at Sarn Gynfelyn, a bank of shingle, stones and sand that extends far out into the sea. The name means Causeway of Cynfelyn – a British king who lived just before the arrival of the Romans.

Sarn Gynfelyn plays an important role in perhaps Wales's most famous coastal legend – the loss of Cantre'r Gwaelod.

The basics of this legend are well known. There once was a large, fertile land that sat in what we now know as Cardigan Bay. It was ruled by King Gwyddno Garanhir and protected from the sea by a wall and a gate manned by his loyal knight, Seithennyn. But one evening Seithennyn got drunk (or distracted by a beautiful young lady depending on who and when the story was told through the ages) and forgot to close the gate. The sea rushed in and Cantre'r Gwaelod was lost.

The morality tale behind the flooding of Cantre'r Gwaelod – of humanity corrupted by drink or lust – has echoes, of course, of the Atlantis myth told by the Greek philosopher Plato sometime around 360 BC.

In Plato's cautionary tale, the founders of Atlantis were half god and half human and they created a utopian civilisation some 9,000 years before his own time. Alas though, the rulers grew increasingly greedy and became morally bankrupt so the gods decided to punish them by triggering a cataclysmic earthquake that destroyed Atlantis and caused it to collapse into the sea.

Plato spent much of his life considering what an ideal society represented and the self-destructive nature of human society. His ideas would heavily influence both Roman philosophy and that of the early Christian church so it's not hard to imagine that the telling of Seithennyn's weakness could have the same moral roots as the original Atlantis tale.

Versions of the Cantre'r Gwaelod disaster have been passed down through Welsh history since *The Black Book of Carmarthen* (a collection of poems and triads written around 1250) first mentioned Maes Gwyddno (Gwyddno's Field). In that reference, the kingdom is flooded because a well-keeper called Mererid forgets to replace the cover of the well. Later versions maintain that the only people to escape Cantre'r Gwaelod did so by using Sarn Gynfelyn to reach the shore.

As with all Welsh legends and history from the Dark Ages, it's often hard to work out where factual characters end and fictional ones begin. No more so than with one of the principal people connected to Cantre'r Gwaelod, and one of the most important Welsh figures connected to the sea – Taliesin, Garanhir's adopted grandson.

There's little doubt that Taliesin was a real person who lived sometime in the sixth century and was highly regarded as a poet throughout the British Isles. He was known to have been the chief bard at the court of three British kings.

Over the centuries, however, his reputation grew to mythic proportions as he came to embody a character that could inspire later generations about the Celtic and Druidic pedigree of ancient Britain.

For medieval antiquarian storytellers, Taliesin was a Zelig of the Dark Ages. He was said to have fought with King Arthur and, in the *Mabinogi*, he was one of the seven remaining knights who returned to Wales with Brân the Blessed's giant head.

The legend of Taliesin will always be associated with the sea because of how he came into being. The story begins inland on the shore of Llyn Tegid (Lake Bala) where a servant boy called Gwion Bach worked for a giant, Tegid Foel, and his wife, Ceridwen, who was a witch.

She was brewing a magic potion to make her son Morfran handsome and wise and Gwion Bach was tasked with stirring the cauldron. Three drops of the potion fell into his mouth as he did so and he was instantly empowered with infinite wisdom and magical powers. He fled, fearing Ceridwen's wrath. Rightly so for she pursued him with a vicious anger. Gwion used his new powers to transform into a rabbit but the witch turned herself into a dog. Gwion countered by becoming a fish and jumping into a river but Ceridwen took the form of an otter and followed him. Next, he turned into a bird but she became a hawk. Finally, Gwion became a single grain of corn. Ceridwen morphed into a hen and ate him.

Only after Ceridwen returned to her normal being did she realise that eating the corn had made her pregnant with Gwion. She planned to kill him at birth but the child was so beautiful that she cast him into the sea in a large bag. Elffin, son of Gwyddno Garanhir, found Gwion washed up on the banks of the Afon Dyfi a few miles north of Aberystwyth. When Elffin opened the bag, he was shocked by the whiteness of the boy's brow and cried out 'tal iesin', meaning 'how radiant his brow is.'

Elffin adopted Taliesin and brought him to live at Cantre'r Gwaelod. One legend maintains that Taliesin was the only person to escape the flood.

Just as the tales of Kenfig Pool and Pennard Castle hark back to real life climate catastrophes, Cantre'r Gwaelod may have its own factual basis regardless of how much it has been embellished and inflated through the centuries. The existence of tree trunks buried in the beach at nearby Borth shows how the sea has swallowed up land that once was home to a large woodland – similar to the sunken forests that can be found in south Wales and on the Gwent Levels.

More fascinating still is the existence of the Gough Map of the British Isles, drawn up in the thirteenth century around the same time that *The Black Book of Carmarthen* was written. It shows two distinct islands located in the heart of Cardigan Bay. Cartography created by the Romans suggests that the Ceredigion coastline may have been some eight miles further west than it lies today. Those two islands might have been mountains back then so, perhaps, Gwyddno Garanhir's kingdom really did exist after all.

9

Farewell to Fairbourne

*Walks from Aberdyfi to Tonfanau
and Llwyngwril to Fairbourne*

The links course on the outskirts of Aberdyfi is one of the signature venues of Welsh golf – a challenging layout nestled just inside the tall sand dunes that offer natural coastal protection from the elements of Cardigan Bay. The present course was built in 1892 not long after the very first round had been played on Aberdyfi Common using flowerpots for holes. One of those original golfers was Bernard Darwin, grandson of naturalist Charles Darwin. He went on to become golf correspondent for *The Times*.

Andy and I are walking on a public right of way that cuts through the heart of the course because, today, we are on a special mission to map a new route for the Wales Coast Path up to Tonfanau, about seven miles up the coast.

Taking the official route would be an easy morning stroll following the golf course and beachfront up to the holiday resort of Tywyn and then continuing across the *morfa* to Tonfanau. The problem with that route (or the coming problem) is that, based on future sea level projections, parts of Tywyn may well not be there in thirty years' time. Instead, a new body of water could intrude perhaps as much as three miles inland – at least as far as the village of Bryncrug and perhaps even further.

Admittedly, some more conservative projections put their faith in increased defences to protect the central parts of Tywyn. The reality though is that both best- and worst-case scenarios are just projections. No one knows for sure what the sea will bring over the coming decades. Whatever the real impact, it's a good bet that anyone wishing to explore the Coast Path will likely find themselves tramping a decent way from what is currently the coast.

I've been contemplating this exercise in psychogeography for a few weeks as the realisation of how profoundly climate change will impact the coast of Wales starts to sink in. Large parts of it are now vulnerable because of sea level rises, seasonal storm surges and the effect of river flooding overpowering estuary systems. In the worst-case scenario, some places could be hit by all three, causing chaos.

I've already witnessed the risks and mitigation actions being undertaken to protect the Gwent Levels and the Cardiff coastal plain. I've seen how surrendering to the sea has changed the landscape on the Llwchwr Estuary and how, in Pembrokeshire, the natural pebble rock defences at Newgale are being breached on a near annual basis.

Remapping the entire Wales Coast Path is a bigger project than I am prepared to take on at this time – partly because it is difficult to emotionally comprehend but also because there are so many variables that will ultimately determine what parts of the coast are defended and protected.

Officially, Natural Resources Wales states that most of the coastline should continue to be protected until the end of the century. The reality, however, is that politics, lobbying, security and environmental sustainability decisions will determine the outcome. Some parts of the most vulnerable coastline – especially around cities and some tourism centres – will be heavily fortified with coast defences. Others – places perhaps with less political clout, less population or obvious economic value – will be sacrificed.

That's why I've decided to use this short section as an experiment. One of today's challenges will be plotting a new route that keeps walkers above the high-water line without intruding on private land. Which explains why, right now, Andy and I are traipsing along a permissive footpath through the heart of the Aberdyfi course, greeting golfers along the way with a friendly nod and taking care not to walk into their line of fire – though judging by some of their tee shots that brings the entire course into play.

The good news for Aberdyfi Golf Club is that the large sand dunes that flank the beach side of the course should provide some protection in the coming years from sea level rises and storm surges. The bad news is that we don't know if these important natural barriers will be enough. The same is true for golf clubs all along the Welsh coast. These links courses find themselves on the frontline of climate change. In some respects, you could call them the canary in the coalmine for the rest of the coast.

Already I'd walked through or near some of the most vulnerable ones.

Near Chepstow, certain holes on The Mathern course at St Pierre, sitting down on the Levels, are boggy at the best of the times. Further along the Hafren Estuary, Peterstone Lakes Golf Club sits by the side of the sea wall and Peterstone Gout. It will likely be breached over time.

As will parts of Royal Porthcawl and Pyle & Kenfig golf clubs, situated close to the water's edge, though both are protected in some ways by natural sand dune systems. Machynys and Ashburnham, two courses on the banks of the Loughor Estuary, surely will require coastal protection if they are to survive.

It's the same story in mid- and north Wales. Borth and Ynyslas Golf Club, almost in touching distance from where I am standing now, doesn't stand a mermaid's chance of surviving unscathed. Most likely it will be swallowed up by the sea as it inundates the sandy headland from the rear, avoiding the reinforced beach defences and flowing through the Ynyslas Nature Reserve.

To the north, Royal St Davids in Harlech, Porthmadog, Pwllheli and many more – all these illustrious golf courses – will be impacted over the next thirty years as the sea rises. Whatever your feelings about golf and the amount of private land it occupies, it's clear that the coming decades of climate change will be devastating for the sport. Those losses will have a knock-on effect on tourism incomes and will signal a change in the way of life all along the coast.

It's not just golfers who should be concerned. A large part of Wales's tourism economy is based near to these links courses and each part will face its own challenges as our climate changes. Sea level rise and intensified storms may yet rob Wales of some of its finest sandy beaches – the ones that are used in tourism brochures to attract visitors from all over the world. How will the iconic three-mile stretch of Rhossili beach be altered in the coming decades? What about Tenby and its dual North and South bucket and spade friendly sands? Here in Aberdyfi, the beach is just a sand wedge away from the golf course. Large parts of it will surely disappear by the end of this century.

Then there are the rocky coves of Pembrokeshire around St Davids – home to the very Welsh adventure sport of coasteering, which involves traversing rocks and cliffs in the sea. How will erosion and increased storms impact its appeal? Likewise the sea kayak trips that launch from the currently sheltered bay of Porthclais. Or the famed surfing beaches like Whitesands, just north of the saint's city? All these activities exist at the whim of the sea, the tides and the weather. They have evolved from our ability to adapt and play in our coastal environment. But that doesn't mean they are guaranteed a future.

In the coming decades, Wales must look to protect the jewels of its coastal tourism industry in the face of increased climate change and volatility. Even as that change attracts more visitors who once would have headed to southern Spain, Italy and Greece before they started to become too hot in the summer months. One EU study, conducted in 2021, found that west Wales could 'benefit' from a fifteen per cent increase in tourism if temperatures keep rising and emissions remain unchecked.

As Wales looks to increase tourism – hoping to replace the employment and revenues lost through the past decades to industrial and rural decline – it also faces another conundrum: the environmental and social impact of the tourists themselves.

Some of that impact comes in the form of carbon miles travelled to reach Wales and from the type of overcrowding already being experienced in parts of west and north Wales. Local authorities in Gwynedd recently opened dedicated rest stops for mobile homes to cut down on the large number of vehicles camping illegally on beaches and mountain roads.

The Wales Coast Path itself is feeling the heat of having to manage large numbers of walkers, notably around what its administrators call 'honeypot sites' – the most popular parts of the path, typically near car parks, beaches and viewpoints. In some places, such as the path down to St Govan's Chapel and the viewing station for The Green Bridge of Wales (both in the east range of Castlemartin), the volume of visitors has eroded the ground down to the bedrock.

Back at the furthest point north on the golf course, a wooden stile points the way for my newly planned route. Except it leads through a rough field of salt marsh that is flooded and hemmed in by wire fencing. This is the very first hurdle in imagining a new inland route for the Wales Coast Path and we've only ventured a mile out of Aberdyfi. Andy and I have experience of scrambling over wire fencing from when we walked through the woodlands of the south Wales valleys. Back then, though, there was a clear track to follow. All I can see now is *morfa* and a very wet future.

Seascape

We backtrack onto the golf course, past three elderly ladies carefully sizing up their putts on the ninth. The stare we receive tells us this is serious business. We hurry along through the sand dunes and return to the beach.

The tailwind from a morning storm still lingers, whipping up the top layer of sand. A ribbon of seaweed and foam deposited by the high tide points our way towards Tywyn. Ahead of us, clear skies over the Tolkienesque topography of the Llŷn Peninsula offer the faint hope of a dry afternoon.

At Tywyn, we are greeted by a plantation of static caravans, stationed just metres from the beachfront. Some owners have built raised benches on their individual plots so they can peer out over the long concrete sea wall.

The wall was constructed back in 1889 by the salt magnate John Corbett, as part of a grand esplanade. He had purchased a large local estate and was keen to fully develop Tywyn as a holiday destination – the town having first attracted English visitors seeking out a picturesque experience nearly a century before. The young Charles Darwin visited Tywyn in the early nineteenth century.

Tywyn's history runs far deeper than the Romantic period. The name means 'sand dunes' and, geologically, the beach area is an extension of the dune complex south of here in Aberdyfi, Borth and Ynyslas along the Dyfi Estuary. Just like Borth, Tywyn once had its own tract of sunken forest.

It also has strong Celtic saint credentials. The town grew from a *clas* (religious settlement) founded in the sixth century by Cadfan, a Breton cleric who is said to have studied at Illtud's seat of learning in south Wales. The church bearing his name is home to a cross-carved stone pillar from the ninth century and is inscribed with the earliest known written example of Old Welsh, the forerunner to the modern Welsh language.

The first historic mention of the stone dates back to 1698 when it stood in the churchyard. At one time it was used as a gatepost. In the 1790s the stone disappeared, only to turn up in a grotto owned by a local gentleman. In the mid-nineteenth century, the stone was returned to the church. Today it is displayed in the nave of St Cadfan's and offers an understated but irreplaceable glimpse into the Wales that was being formed back in the Dark Ages.

As for Cadfan, life on the mainland proved too hectic and, like many of the most devout Celtic saints, he sought solitude out in the ocean, in his case on the island of Ynys Enlli (Bardsey) some two miles off the tip of the Llŷn Peninsula. Cadfan built a monastery there and, over the centuries, it became recognised as one of the holiest sites in all of Britain and Europe.

No one knows what attracted Cadfan to Enlli (apart from its remoteness presumably). But it's not inconceivable that he would have heard tales about a contemporary warrior who was gaining fame battling the Saxon invaders across Britain. His name was King Arthur and his exploits, as we know, would become the stuff of legend. So much so that, as bards and storytellers through the centuries embellished and burnished his reputation, it is now impossible to know what was fact, and what was fiction.

Two tales of Arthur and his knights are closely associated with Ynys Enlli. The first is that the magician Myrddin (Merlin) is buried on the island. The second, even bolder, claim is that Arthur himself retired there to heal his wounds from battle and that Ynys Enlli is the fabled island of Avalon. There, Arthur's magical sword, Excalibur, was forged.

It's a great story, the veracity of which is only undercut by the fact that many other parts of Wales, southern England and Scotland also claim to be Avalon.

Tywyn's modern popularity as a tourist destination grew in the 1840s when a grand boarding house, Neptune Hall, was built on the seafront. Today, that name has been assumed by the local caravan park that sprawls off across the fields behind the beach.

More than 3,000 people live in the town inland of the sea wall while another 4,000 reside in the greater Tywyn area – much of it on the flanks of the wide Afon Dysynni Estuary.

Here, next to the caravan park, my mini remapping experiment begins in earnest. Instead of following the Coast Path along the beach we turn inland up Neptune Road until we arrive at Faenol Isaf, a winding street of suburban houses anchoring a residential housing estate. We stroll past residents tending to their gardens and going about their daily business until we reach an alleyway. My map indicates this is a public footpath leading northeast away from the sea. It's just the route we are looking for because, based on the less optimistic flood and sea level projections, all of Faenol Isaf and Neptune Road will be underwater by 2060 unless there is a significant increase in what already looks like a very expensive and expansive protection scheme.

The footpath runs past someone's garage wall before vanishing into an overgrown clump of stinging nettles. It doesn't fill us with confidence but we follow it, doing our best to avoid getting stung in the process. We emerge into open fields. Sheep graze nearby. The long fairways of Aberdyfi golf course lie off in the distance.

The next bit involves crossing the main coastal railway line via a pedestrian access gate, skirting around another static caravan park, crossing the main road into Tywyn and then weaving our way behind a group of houses that must have been built after this right of way was first established.

'Didn't we just walk through someone's back garden?' asks Andy once we finally emerge onto a country lane on the northern outskirts of Tywyn. 'How do you think they're going to feel when your new Coast Path route bisects their lawn?' he adds.

At least we are still headed in the right direction – northeast, keeping on the high ground above the salt marsh. On the map, I spot another footpath that follows the Talyllyn narrow gauge railway up the valley away from the coast. This seems like a good sign so we follow the track inland.

The railway first opened in the mid-nineteenth century to carry slate down from local mountain quarries. When that trade declined, it was reinvented as a tourist train. In the 1950s, a young enthusiast volunteered as a guard on the Talyllyn line. His name was the Reverend W. Awdry. Later, he would turn his passion into a series of books called The Railway Series which gave birth to Thomas the Tank Engine and many other locomotive characters. Awdry based one of the railway lines, Skarloey, on Talyllyn. In the books steam train characters such as grumpy Duncan and Sir Handel ride the Skarloey Railway, which is owned by Sir Topham Hatt – known to generations of children as 'the fat controller'. You'd never get away with that nickname today.

We wander through farmers' fields, keeping the railway to our left. The rain has returned and the ground under our feet squelches even though this is the height of summer. If you are prepared to ignore the more dire projections for annual storm surges, Bryncrug village is where I've estimated the rising sea waters will run out of energy.

A single-track country lane leads us back towards the coast. Below us the Afon Dysynni Estuary flows through the salt marsh to a large lake. The map says it is called Broad Water but it must have once had a Welsh name. We pass old farmhouses with fine views over the low land and the town of Tywyn. One, Llechlwyd, has enough land running down into the salt marsh that, one day soon, it might be advertised as beachfront property. Everything below us – including the existing footbridge and main railway bridge – will soon be at the mercy of the rising sea.

Tonfanau is our final destination for the day. From here we can catch the Welsh coast mainline train back to Aberdyfi. It is a particularly windswept and isolated place, not somewhere anyone would choose to spend a great deal of their time. Which makes it all the more incredible that, back in 1972, the UK government chose Tonfanau as a temporary relocation settlement for one thousand Ugandan Asian refugees who had been expelled from their country by the dictator Idi Amin.

The refugees had been given just ninety days to leave Uganda even though many of the families had lived in East Africa since the late nineteenth century.

Many of the Ugandan Asians held British passports and around 28,000 headed to the UK. They were allowed to leave with just £55 (about £500 today) and two suitcases. Their arrival caused a political furore not dissimilar to the current arguments over illegal immigrants arriving by boat across the English Channel.

It also presented a logistical headache for the Conservative government of the day. Where could such a large number of people be housed on arrival? The decision was taken to establish temporary settlement and housing camps for the Ugandans in old military bases around the country.

Some were close to urban centres but others like Tonfanau were in the middle of nowhere. Still, it was deemed suitable for 'acclimatising' the refugees into British life. One look at this barren stretch of coast told me everything I needed to know about the culture and climate shock these émigrés must have experienced, having been bussed from Heathrow airport directly to this edge of Gwynedd.

Some of the local community were very welcoming to the new arrivals but they also had to endure hostility and racism – rural mid-Wales didn't exactly have much experience of being a cosmopolitan melting pot. And yet, in the years that followed, the Ugandan Asians integrated into British life, injecting an entrepreneurial energy and becoming one of the most successful immigration stories ever seen. They also had the last laugh over Idi Amin. What he had neglected to understand was that they accounted for about ninety per cent of his country's tax revenues. When he kicked them out, the Ugandan economy all but collapsed.

As we wait on the platform for the train back to Aberdyfi to arrive, I try and imagine what those new arrivals must have felt experiencing the Welsh weather for the first time. Twenty-four hours before they would have been soaked in the humid heat of Uganda. They then arrived in an undoubtably wet Welsh winter. It would have been a shock to their system and surely put a chill through their already dampened spirits.

One of the climate change effects few of us have given any thought to is how permanently altered weather patterns will impact our own mental well-being. Already Wales, along with the rest of the European Atlantic coast, is experiencing increased storm systems that bring days and days of continuous rain, waterlogging not just our fields but also our minds. People living in southern Europe, meanwhile, must contend with months of drought and temperatures so high that they pose a risk to health. Other parts of Asia, the Middle East and Africa may soon be uninhabitable because of the heat. All of these climate changes also change our psyche, our sense of comfort in our home and our well-being as a society. Climate change is set to be a mental health emergency as much as it is a crisis of physical displacement and natural disasters.

One of the first things that strikes you as look down at Fairbourne from the slopes of Pen y Garn is just how neatly and precisely planned the physical footprint of the village is. Most Welsh villages (even seaside ones) are a jumble of winding streets, often veering steeply up the side of a hill or round a cliff face. Not Fairbourne. It looks more like the type of grid layout you might find in a US suburb. There is one main tributary – Beach Road – that connects with other smaller residential streets. On one side of Beach Road, a miniature steam railway runs down to the sea wall before turning north and puffing its way (very slowly) to Barmouth Ferry station on the outskirts of town. In the distance, we see the iconic wooden viaduct and footbridge that spans the Afon Mawddach and connects to neighbouring Y Bermo (Barmouth).

Seascape

Our day begins in the village of Llwyngwril about four miles to the south. For some reason best known to the locals, it is festooned with knitted creations made by local 'yarn bombers'. The woollen centrepiece is Gwril the Giant. He peers out at passers-by from an old stone bridge in the centre of the village. According to local legend, Gwril was a lowland giant. His cousin was a giant named Idris who ruled a nearby mountain from his chair. The peak, Cadair Idris, is named after him and legend maintains the two giants liked to throw rocks at each other for sport.

We escape before Idris finds his range this morning but are soon in the cross hairs of a greater power – a squall that bears down on us from across the bay at the tip of the Llŷn Peninsula.

Almost as quickly as it starts, the storm is gone. The sun appears and, below, the coastline and the town of Fairbourne come into view.

In some ways, Fairbourne resembles Tywyn in terms of its modern development. It was first developed in 1865 by the Cardiff bus and tram entrepreneur Solomon Andrews. He purchased the land and built the first properties while also drawing on his transportation expertise to build a horse-drawn tramway that connected the nearest railway line to the coast. The tram route is now the miniature railway that runs along the promenade. At the time, this land was still known by its Welsh name, Morfa Henddol (named after the two rivers that converge there). But in 1895 an English flour baron called Arthur McDougall bought up a chunk of the land, turned it into a seaside resort and renamed it Fairbourne.

That's where the similarity ends. While Tywyn's defences are being bolstered, Fairbourne has been handed a death sentence. In 2014, Gwynedd Council declared the town will no longer be habitable from 2054, because of a combination of sea level rise, severe storms caused by climate change and the geological conditions surrounding the village. Fairbourne faces a triple flooding threat from the sea, the Mawddach Estuary and the river run-off from the surrounding hills.

Back in 2017, Natural Resources Wales spent £6.8 million on nearly two miles of reinforced concrete sea defences to protect more than 400 properties. But defending Fairbourne will only get harder over time. Hence the long-term decision to 'walk away', in the parlance of government. The council has outlined a plan to essentially 'decommission' Fairbourne by tearing down homes, removing its roads and letting the sea turn the land back to salt marsh. Perhaps, at that point, it will revert to its original and accurate Welsh name.

The decision to abandon Fairbourne, to protect Tywyn, to double down on Cardiff – in fact every climate change protection decision in Wales – depends on a complicated set of financial, economic, commercial, management, social and environmental factors established by the Welsh government's Flood and Coastal Erosion Risk Management Programme. It in turn is influenced by the UK Treasury's Green Book – the government's equally complicated bible for how to appraise policies, programmes and projects.

In its most simple explanation (if that is possible) the programme takes into account a series of critical success factors that have to be met in order for flood protection action to be taken. They include: the need for the proposed solution to complement local flood risk strategy; to offer value for money; to match the capabilities and capacity of the companies that would supply the services; to be able to be funded from available budget and, frankly, to be physically achievable and maintainable.

Based on these (and other business and governance criteria), decisions can be made about how to defend the coast from the sea. There are four options outlined for those decision makers.

The first is to 'walk away' – the fate facing Fairbourne. The second is 'business as usual', which means maintaining the current level of investment and support. The third offers a 'minimum of two intermediate options' to escalate coastal defences. The final is 'do maximum', which means to invest what is needed to protect local communities and businesses.

Local residents, not surprisingly, feel abandoned by the decision to walk away from Fairbourne. Some have campaigned for the government to take a second look at how the village could be saved, even suggesting the installation of 100 tetrapods, concrete structures that can reduce the force of incoming waves and are commonplace along the shores of Japan. Most of all though, the residents feel they have been forgotten.

Fairbourne might just be one small village of 1,200 people, and it may well be undefendable against the power of the sea. But it highlights how unprepared most of us are to acknowledge the reality of climate change relocations.

Where will the people of Fairbourne go? With thirty years warning, it's likely that the abandonment of the village will happen gradually with families migrating into other local areas or venturing further afield. But other communities might not have so much time to adapt and relocate.

Wales, of course, is a bit part in the global climate migration story. But the very fact that thousands of Welsh people might have to move at short notice in the coming decades shows just how big a change society is about to experience as millions more people in climate vulnerable regions are displaced.

Many thousands of international climate refugees will keep seeking safe haven in the UK and the coming decades will bring far greater migration challenges than the ones we face today. It will demand a response from all of us that is far more sympathetic and accommodating than the current mood of so-called developed societies. It will require an element of humility and a dose of historical education to demonstrate how open borders can build stronger, more successful and kinder societies – and have done in the past. The success of Uganda's Asian population in the UK is just one example. If the fate of Fairbourne shows us anything, it is that we all now are vulnerable to the forces of climate change. And that we can take nothing for granted.

We walk to the seafront and then along the sea wall following the miniature train line. Some of the houses along Fairbourne front are neatly kept but others feel already abandoned. Some empty lots are overgrown. In fact, the whole town has a sense of finality about it – surely not what Arthur McDougall would have imagined back at the start of the twentieth century. It was his family that first created what we now know as self-raising flour. If only he could have applied the same wizardry to Fairbourne.

10

The Town That was Built on a Beach

Walks from Harlech to Porthmadog and Borth y Gest to Criccieth

Even if you've never visited Harlech Castle, you'd probably recognise it if you saw it. Its unique selling point is its position, perched high on a promontory overlooking an expanse of sand dunes that stretch for one and a half miles before reaching the sea. The castle's location is what makes it one of the most photogenic of all Welsh historical attractions. It regularly graces the covers of travel magazines and tourism brochures.

Given this strategically smart location, you'd be forgiven for thinking that Harlech was built by the Normans in the thirteenth century to keep a watch for attackers approaching from the coast. By the time they'd crossed the dunes, the castle's garrison would have mobilised and be ready for battle.

Except, when Harlech Castle was constructed, there were no sand dunes. The castle walls abutted the sea because the main threat lay with the native Welsh population located inland who the Normans were trying to suppress. They even constructed a stone path of 108 steps that rose up the cliff face and allowed defenders to be resupplied by ship when the castle came under siege. It was called 'The way from the sea'.

Just a few decades after Harlech Castle was completed, the sea adjacent to the cliffs started to silt up with tiny particles of rock and sand. They were carried by waves and powerful south-westerly winds that, over time, created a vast sand bank to the northwest of the castle. Where the coastline once met the mouth of the Afon Dwyryd, the sand formed a finger-like spit driving some of the estuary water inland where it formed a lagoon. Centuries later, that lagoon also silted up with sand and mud creating a vast salt marsh known locally as Morfa Harlech.

Today, I'm going to walk from the sand dunes at Harlech up to Porthmadog. It's a thirteen-mile journey and will take about six hours. I've taken on fuel in the form of coffee and a bacon and egg bap and now, after one final glance up at the impressive Harlech Castle, I set off. Looking at my map, I can already tell I'm unlikely to run into many other hikers today. Ahead of me are long stretches of desolate marshland that make up Morfa Harlech. As I study the route, I'm reminded of Geoff Jones's point about walking the Pembrokeshire Coast Path – you've got to walk at least an hour away from the nearest town or village to find solitude. That's not going to be a problem on this walk.

Wild grasses and reeds line the side of the path as it criss-crosses the wetlands. An odd harmony of birdsong flows out of the bushes but, as the foliage fades and the path enters a field of buttercups, even the wildlife sounds fade away.

Two farmhouses – Glan y Môr and Glan y Morfa – sit on the edge of Morfa Harlech Nature Reserve. For some reason best known to the Coast Path's mappers, the route runs straight through the front garden of the second house. A front garden where a very large, unleashed mixed-breed dog – part wolf I'd wager – is sitting. The door to the house is open so I call out to let the owners know I'm about to walk across their land. No one replies. The dog, however, has seen enough. It has jumped to attention and is barking with considered hostility. I retreat with haste, closing the gate as the dog gives me a final snarl of satisfaction. An alternative route through a churned-up, muddy lane, clearly used for herding livestock, suddenly seems quite appealing. I keep to the sides, avoiding the slushier parts, before hopping over a metal fence to rejoin the official path. The dog watches me every step of the way.

The Morfa Harlech sand dunes took centuries to fully form and continue to evolve. These naturally dynamic systems are constantly changing in response to changes in rain, wind and sand supply. They are similar in character to the dunescapes that formed in south Wales and swallowed Kenfig Castle (perhaps Harlech should consider itself lucky to have been built on a cliff). Yet despite their prominence along sections of the Welsh coast, dunes can only form under certain environmental and topographical circumstances and so cover just 0.3 per cent of the country.

Dunes first grow when sand builds up around an obstruction such as seaweed or debris on the shore. Once the embryo dune starts developing, specially adapted plants like marram grass take hold. This grass is deep rooted and can grow rapidly to keep pace with sand that might smother it. Further inland, established dunes become more vegetated with other grasses and plants that have adapted to the dry soil and salty atmosphere.

During much of the last century, dunes were seen as fragile environments that required stabilising. Large amounts of marram grass and even fast-growing conifer trees were planted on them; sand was fenced in and people kept out. The result of this dense vegetation growth was that the dunes became too solid and rooted and nearly ninety per cent of the open sand disappeared, destroying habitats for species like the fen orchid and the crucifix ground beetle.

Nowadays ecologists better understand that dunes are mobile ecosystems, where sand needs to move through the landscape and create new areas that can be colonised by a wide range of specialist species.

Which is where a dune protection scheme called Sands of Life played an important role. From 2018 to 2024 it worked to restore over 2,400 hectares of sand dunes across ten separate sites along the Welsh coast including the dune ecosystems I've witnessed so far at Kenfig, Merthyr Mawr and here at Morfa Harlech. Sands of Life let nature do its thing – allowing the natural generative processes created by wind-blown sand to take place.

Understanding the importance of sand dunes is particularly important because, while large parts of the Welsh coast are increasingly vulnerable to climate change, the dune complexes may actually expand as powerful tides shift more sand onto certain beaches.

The other impacts of climate change on these fragile ecosystems are less clear. We know that sand dunes respond quickly to changes in environmental conditions. Warmer winters and wetter summers will increase the growth of coarse vegetation, and the increasing severity of winter storms may improve the mobility of the dunes but they can also cause erosion. Nobody knows for sure but, due to their naturally dynamic nature, dunes may actually be able to cope with changes like sea level rise and increasing storminess without being destroyed, by migrating further inland or changing position on the coastline.

Gulls ride the thermals above a large round hill called Ogof Foel. They are my only company this morning and I feel privileged to have Morfa Harlech all to myself.

A wooden sign points to higher ground leading away from the salt marsh. From this vantage point, the entire Afon Dwyryd Estuary comes into view. The sharp peaked ridges of Eryri National Park line the horizon – thick white clouds hovering over them. Dense woodland covers the undulating slopes that lie between the mountain and the estuary. It is a quintessentially Welsh vista apart from one anomaly – the brightly painted white, pink, yellow and blue buildings, seemingly transported straight out of Renaissance Italy, that nestle amongst wooded hills on the other side of the estuary. This is Portmeirion, the theme village created by aristocrat architect Clough Williams-Ellis over a fifty-year period starting back in 1925. It's one of the most unlikely sights in all of Wales and, in my opinion, one of its most appealing.

Williams-Ellis knew he wanted to be an architect from the age of five and, more than anything, he wanted to build his own village. It helped that he hailed from one of the oldest and wealthiest land-owning families in north Wales; he could claim a direct lineage to Owain Gwynedd, Prince of Wales in the twelfth century.

The peninsula that he would later develop belonged to his uncle. Back then, it was known as Aber Iâ and consisted of a dilapidated old mansion lived in by an eccentric old lady tenant. No improvements had been made to the ramshackle estate in decades. When she died, the uncle asked the young architect if he knew of anyone who might rent or even buy Aber Iâ. Williams-Ellis had never set foot on the estate even though he only lived five miles away. In fact, he'd only ever viewed it from across the estuary just as I was looking at it right now. But that glimpse had been enough to fascinate him. He loved what he called 'the romantic aspect' of Aber Iâ. He took a visit to the estate and immediately knew it was where he wanted to build his model village.

Williams-Ellis first set about adapting the old mansion house down by the water's edge and then, over the years, created what he called his 'architecture of pleasure', an Italianate medley of brightly painted houses, cottages and landscaped gardens perhaps loosely modelled on Portofino (though Williams-Ellis always rejected this, despite his open admiration for the Genoan coastal town).

He referred to his vision as light opera and his approach to development harked back to the Picturesque estates of the late eighteenth century such as Hafod in mid-Wales, which had been designed to complement the natural beauty of their surroundings.

'I wanted to show that you could develop a very beautiful place without thereby defiling it. If you took enough trouble with enough loving care and knowledge you could even enhance what God had given you as a site to start with,' he told the BBC back in 1969, when Portmeirion had already succeeded in capturing the public's imagination and attracted the interest of numerous film and TV location scouts, including ones for *The Prisoner*.

With the tide out, the bright yellow sands of Traeth Bach stretch almost the entirety of the estuary. A couple of walkers venture out to Ynys Gifftan rock in the middle of the estuary – perhaps trying to recreate the famous scene from *The Prisoner* where Patrick McGoohan sprints along the sands in a desperate attempt to escape his mysterious captors. I hope they've timed their hike correctly because, when the tide turns, this estuary rapidly floods. I also hope they are more careful with their footing than McGoohan. He sprained his ankle while filming that scene and it had to be finished using a stunt double.

The Coast Path follows the raised sea wall built to protect farmland and the few houses dotted on the edge of the salt marsh. It is reinforced with huge slabs of slate. A few sheep graze on the *morfa*, taking advantage of the nutrient rich grass before the tide rushes back in.

Finally, after a two-hour walk around the estuary, I arrive at the entrance to Portmeirion and wander down into the pedestrianised village, through the faded salmon Gatehouse arch that is shaped to complement the cliff face, past a kaleidoscope of yellow, blue, red and white façades and into the main piazza.

Portmeirion might easily have been dismissed as the whimsical indulgence of a very wealthy man but Clough Williams-Ellis built the village for others to visit and learn from – not as his private folly. He wanted to show that nature and architecture could go hand in hand and he also was committed to reimagining things that others were keen to discard. A number of the buildings in the village were rescued from demolition in other parts of north Wales and then integrated into his architectural vision.

He railed against urban development that was eating into the countryside and he was instrumental in the creation of the National Parks and Access to the Countryside Act of 1949, which created the framework for Eryri and Bannau Brycheiniog.

Given Williams-Ellis's pedigree and embrace of nature, it's not surprising that when the famous American architect Frank Lloyd Wright made his only trip to Wales in 1956 he visited Portmeirion.

Wright's grandparents came from Taliesin, the village named after the famous bard and child of the sea, just north of Borth. He named his Wisconsin home Taliesin and his Arizona winter retreat, Taliesin West. In some ways, he and Williams-Ellis were kindred spirits. Both Taliesin designs incorporate and integrate the built environment with the natural world – though in a far more understated and minimalist design than Portmeirion.

The town of Porthmadog is about as far removed from Clough Williams-Ellis and Frank Lloyd Wright's vision of building in harmony with nature as you could imagine.

I've left Portmeirion and walked back to the point on the peninsula where the Afon Dwyryd and Afon Glaslyn meet. Here, on high ground, the town of Porthmadog comes into view. A long causeway, known as the Cob, spans the mouth of the Glaslyn, connecting this side of the river to the town. It is an impressive feat of engineering – perhaps as great an attempt to tame the sea as I've seen so far along the coast of Wales.

The Cob was conceived by William Madocks, a local landowner, and built at no small expense to him between 1808 and 1811. Madocks was a Member of Parliament for Boston in England but his family connections to north Wales dated back to Norman times. He was educated at Oxford and, like many young men coming of age during the Industrial Revolution, was fascinated with engineering and the possibilities of how his generation might shape this fast-changing world.

In 1798, Madocks bought the Tan-yr-Allt estate, situated on the western bank of Traeth Mawr and quickly set about reclaiming the surrounding salt marsh for agricultural use. He constructed a semi-circular sand embankment that ran parallel to Afon Glaslyn for two miles and reclaimed 1,000 acres of land.

Following the 1800 Acts of Union between Britain and Ireland, the brightest minds were looking for new speedy transport routes to connect the two lands. Madocks wanted to establish a ferry route to Ireland at Porthdinllaen further up the Llŷn Peninsula. A faster way of getting from mid-Wales to the Llŷn was needed if this was to be effective. That's when he came up with his grand plan – to build a second embankment that would bridge the estuary.

In 1807, he succeeded in getting passed an Act of Parliament that allowed him to start construction. The catch was that he alone would be responsible for its cost. In return, he would receive the 6,000 acres of land reclaimed from the sands and some of the rent from leasing out the new farmland.

Madocks's engineer on the first project had died during the time it took to get the go-ahead for the new project. So Madocks took the unorthodox step of asking his gardener, John Williams, to oversee the Cob.

Williams had some experience working on the construction of the first embankment but it was still a bold selection. However he was a Welsh speaker which helped with managing the local workforce. It was Williams who identified the need to build a barrier over nineteen feet (six metres) high and incorporate five sluice gates to allow the Glaslyn to flow into the sea.

Some 300 men were hired to work on the project – most of whom also had to be housed and fed. Madocks was burning through his funds but hoped the project would be completed by May 1809. However, in ways that mirror the problems Sir John Hawkshaw would later face with the Severn Tunnel, construction was hampered by the power of the tides flowing in and out of the estuary.

By 1811 the Cob was finished but Madocks was severely in debt and being pursued by his creditors in the courts. Then, disaster struck. In February 1812, a winter storm and high tides broke through the Cob. The project had already cost Madocks some £60,000 and he was effectively bankrupt.

It was at this point that the local community stepped in to help. Initially, many of Madocks's fellow landowners had been against the embankment but now, having seen the benefit of improved transportation links (and having been alerted by Williams to Madocks's problems) they rallied around and raised the funds needed to repair the breach.

Despite the community bailout, Madocks had exhausted all his funds and lost nearly all his possessions in his Canutian pursuit. Yet reclaiming that much land had one important, unintended consequence. By diverting the Afon Glaslyn, Madocks had created a new natural harbour on the west side of the estuary that was deep enough for ocean-bound ships – just at the time when the local slate mining industry was starting to grow.

Port Madoc – named after Madocks – was built on the sand banks and salt marsh next to the new harbour. It opened in 1825 and soon was exporting large amounts of slate transported from quarries up the valley on the Ffestiniog Railway via the Cob. By the 1870s more than a thousand ships were sailing into and out of the port, carrying the thousands upon thousands of tonnes of slate that would provide the roofing for new towns and cities all over the world. Over time, the town that grew up around the docks would become known as Porthmadog.

It is late in the day and I've been walking since 10 a.m. I am tired and a little chilled now that the sun has started to set but, as I cross the Cob via a low footpath, I linger a little, taking the time to appreciate the Traeth Glaslyn Nature Reserve on the inland side of the estuary embankment. A family of ducks zigzag single file up the water channel in front of me.

Ahead lies Porthmadog harbour. Today, despite the fact that the boom days of the slate industry are long gone, the town remains home to more than 4,000 people. Most of them live close to this old part of the town that, essentially, is built on a beach. Only Madocks's Cob stands in the way of them and the rising sea.

Yesterday's tramp from Harlech to Porthmadog was rewarding but frankly exhausting. So my plan today is to do an easy walk – just six miles along the Llŷn Peninsula coast from Borth y Gest (one bay over from Porthmadog) to the castle at Criccieth. I watch as the tide in the bay recedes at a rate that makes me think perhaps someone has pulled the plug on the entire Glaslyn Estuary. The small boats bobbing in the water become statues as they are stranded on the sandbank.

This section of the Coast Path has a very different feel to the wild desolation of yesterday. Expensive homes with names like Heather Ridge, Fuchsia Cottage and The Haven look down on me as I walk. The names seem more than a little out of place here on the Llŷn Peninsula. This, after all, is considered one of the heartlands of Y Fro Gymraeg (Welsh-speaking Wales) where over seventy per cent of the 20,000 population speak Welsh as their first language.

Seascape

A large jellyfish lies stranded on the beach directly below Porthmadog Golf Course. Two kids circle around it – one pokes it tentatively with a piece of driftwood. Buried in the sand are huge slabs of micro-layered black and grey slate. They look like stranded giants – maybe this is what inspired the tale of Brân the Blessed's severed head.

At the furthest point of the headland sits a solitary white cottage. It's known as the Powder House because, a couple of hundred years ago, it was used to store the explosives needed for the nearby slate mines. Understandably, the local population was concerned about ships full of gunpowder docking at Porthmadog so the cargo was unloaded here and carried by horse and cart to the mines.

Around the headland I walk onto Traeth Morfa Bychan (Black Rock Sands). A cluster of gulls peck at seaweed and some unfortunate beached crabs close to this morning's high tide line. Black Rock is a rarity, not just because of its natural beauty (which is impressive), but because it's one of the few beaches in Wales where vehicles are permitted to drive. Having walked so much of the coastline over the past months, it's a shock to see camper vans and cars parked on the sand. It feels like an intrusion, a blight almost. At the furthest end of the bay, a VW Golf is parked down by the water's edge. I will it to get stranded like the crabs. I have a hunch that the gulls would take the car apart in no time.

From the top of Graig Ddu (Black Rock), I can see Criccieth Castle, standing broken but still proud on the cliffs in the centre of the town. Unlike many of Gwynedd's castles, this one was first built in the eleventh century by the Welsh king Llywelyn the Great, as a demonstration of his power and authority over rival Welsh princes throughout Wales.

When Llywelyn's grandson, Llywelyn the Last, was defeated in 1282 by Edward I, the new English ruler shored up the castle to repulse repeated Welsh attacks. It wasn't until Owain Glyndŵr's rebellion against the English crown that the castle was compromised. He sacked it in 1404, leaving the ruin we see today.

The artist J. M. W. Turner painted the castle on a tour through Wales in 1835. His watercolour captures a great storm rolling in and breaking over the castle. In the foreground, local people gather to retrieve items being washed up by the sea.

The path follows the train line into town arriving at the beachfront that Turner painted all those years ago. There's no need for a salvage operation today. It's a sunny early September day and Criccieth has that end-of-season beach resort feel to it. The school summer holidays are over, so it's mainly retirees who wander the small square nestled under the cliffs on the east side of the castle and queue for ice cream at Cadwaladers a little further up the hill.

Just under a century ago this area was known as Abermarchnad and home to ten beachfront cottages. Then, on the night of 27 October 1927, a violent storm – combined with a potent spring tide – hammered into the Llŷn Peninsula. The storm surge and waves smashed down the walls of the cottages with such ferocity that it was decided never to rebuild there.

On the other side of the castle lies the west beach esplanade. From here I can see just how exposed this town is to the power of the sea. The 1927 storm was considered a once in a lifetime event but Criccieth has been battling the elements for centuries. It sits at a particularly exposed location on the Welsh coast where the fetch of the waves (the distance they can carry without meeting any obstacles) reaches nearly 5,000 miles out across the Atlantic Ocean. Those waves generate and deliver an enormous amount of energy by the time they hit the shore. Below me, rows of groynes have been positioned along the beach. Doing so has altered not just the appearance of the beach – the sand and shingle on the side of one groyne is almost one and a half metres higher than the other side. It has also accelerated the erosion of the cliffs, threatening the Coast Path, and starved the town's eastern beach of sand. This, in turn, threatens further deterioration of the cliffs closer to the town.

Seascape

Harlech, Porthmadog, Criccieth. All three of these neighbouring coastal towns will have to adapt and possibly relocate some of their inhabitants in the coming decades as different forces of climate change eat away at the coast, inundate it with sand or flood low-lying land. Just as Fairbourne must come to terms with surrendering itself to the sea, many more towns and villages will have to contemplate the need to withdraw from the coast, and how to do so in the least damaging way – both for the people forced to relocate, and the communities where they might now have to settle and integrate into. In technocratic talk, this is called a 'managed retreat'. You can see why the concept is so difficult to comprehend.

In some instances, it will simply mean building on higher ground close to existing but vulnerable towns and cities. A big part of our present city infrastructure crisis can be attributed to where developers have built over the past century. All too often, construction projects undermine the coastal land being developed – extracting groundwater and compromising the natural balance of the land. As a result, that land starts sinking. Bangkok in Thailand is particularly at risk from falling into the sea, as are Lagos in Nigeria and Dhaka in Bangladesh. Indonesia's capital Jakarta, home to thirty million people, is sinking so quickly that government offices are being relocated to higher ground.

Adaptation might also mean developing new urban centres inland or reinvesting in upland areas that once were important towns but have declined in recent years. Here in Wales, the hill towns of Merthyr Tydfil and Aberdare once dwarfed Cardiff and Newport both in size and their economic importance. As the sea encroaches further into the Gwent Levels, it could be that people choose to adapt by moving back up the valleys to these towns.

Or what about the town of Machynlleth, located about ten miles up the Dyfi Estuary from Cardigan Bay? It was where Owain Glyndŵr established his parliament in the early fifteenth century and many people still consider it a de facto capital of

Wales because of its location in the middle of the country. Perhaps the coming decades will see Machynlleth grow once more in strategic importance and as a destination for climate migrants.

How we build in the places we relocate to (or choose to defend for that matter) will determine the sustainability of those decisions. The sea is teaching us now, in real time, that even our strongest artificial structures – the best sea walls and barrages for example – can be compromised when they fail to integrate with the way nature works. Even iconic examples of human design standing up to the sea, such as the reinforced front façades of the grand townhouses on St Malo seafront in France, will surely be vulnerable in the decades to come.

Our future homes, buildings and infrastructure will have to be designed in ways that respect the power that the sea and the weather delivers with increasing force and unpredictability. That's not just an issue for towns and cities on the coast. Water flows through the path of least resistance and much of that water will flow down from the hills in the form of rainfall. Knowing where to build and not (on estuary floodplains for example) will be critical. As will the design and materials we use.

Even here, the sea can provide us with inspiration if we know where to look. Take the example of the 180-metre-high Gherkin tower in London. Its core design was inspired by a marine animal, the Venus's flower basket sea sponge (*Euplectella aspergillum*), which has a particularly strong lattice-like skeleton that allows it to survive more than half a mile below the sea surface. The external scaffold of the Gherkin mimics the cage structure of *Euplectella aspergillum*. On land, this design allows winds to pass easily around the building and forces air upwards through it, reducing the need for air conditioning by fifty per cent.

More ambitious still is the concept of 'seasteading' – establishing standalone communities and cities that float on the ocean, untethered to land and powered (in some concepts) by thermal currents. If this sounds a little too futuristic and perhaps apocalyptic, then maybe that's because the initial ideas stemmed from the libertarian thinking of Silicon Valley and some of the same anti-establishment motivations that have prompted tech billionaires to buy themselves end-of-the-world estates in New Zealand and Patagonia, or pursue projects to colonise Mars.

That said, who am I to discount the possibility of living on or even under the ocean if sea levels continue to encroach on the places we currently call home. Already, some tangible floating city projects are in development. In South Korea, the city of Busan is building OCEANIX, a community of three interconnected floating platforms, totalling fifteen and a half acres. These platforms sit on the coast and are connected to the mainland by span bridges which create a lagoon. OCEANIX will initially be home to 12,000 people but could house over 100,000 in the future.

No matter how or where we build in the future, our architects and engineers will have to work with the natural world rather than trying to tame it. As Frank Lloyd Wright advised his students: 'Study nature, love nature, stay close to nature; it will never fail you.'

11

Whose Home is This Anyway?

Walks from Pwllheli to Abersoch and Porthdinllaen to Trefor

On 8 September 1936 three men – Saunders Lewis, Lewis Valentine and David John (D. J.) Williams – set fire to a newly constructed RAF training camp at Penyberth, half a mile inland from the beach at Carreg-Y-Defaid, and within a very big stone's throw of the Coast Path running between Pwllheli and Abersoch.

This was no ordinary act of arson and these men were no ordinary vandals. All three were founding members of Plaid Genedlaethol Cymru (the National Party of Wales). They were also respected members of the community. Lewis was an academic and poet, Valentine a leading Baptist minister and D. J. Williams a talented short story writer. The fire they set was an act of protest. It quickly became known as *Tân yn Llŷn* (Fire in Llŷn). It would also provide the touchpaper for what we now know as modern Welsh nationalism.

The UK government had chosen Penyberth as a military base

only after two other proposed sites in England had been blocked by local protests. There were similar objections here on the Llŷn – in fact over 500,000 people across Wales signed a petition against the plan. But the Prime Minister, Stanley Baldwin, pressed ahead. Saunders Lewis was particularly vocal in decrying the decision – one that had been taken exactly 400 years after England passed the Acts of Union that annexed and integrated Wales. He railed against Baldwin's government accusing it of turning one of the 'essential homes of Welsh culture, idiom, and literature' into a training ground for barbarism.

The three men openly claimed responsibility for the fire and were put on trial first in nearby Caernarfon where the jury failed to reach a verdict. Their retrial took place at the Old Bailey in London. There, they were convicted and sentenced to nine months in jail. When they were released, some 15,000 people cheered their arrival back in Caernarfon.

The second of my walks on the Llŷn Peninsula begins in Pwllheli early on a Tuesday morning, meandering along the parade at the town's south beach. My destination is Abersoch, an old fishing village about eight miles down the coast that in recent decades has seen a large influx of second homeowners – a growing trend throughout parts of coastal north Wales. This is creating tension with locals who complain they are being priced out of the housing market – especially young people who are having to move away from their home towns and villages because of price inflation. In Abersoch, nearly half of all available housing is owned by second homeowners. Some flats are selling for upwards of £1.5 million. Even some beach huts are going for more than £200,000. Here in Pwllheli, the sentiment is uncompromising. A notable number of stickers posted on lamp posts and street signs display the following message: 'Wales is not for sale.'

It was also here in Pwllheli in 1925 that Lewis, Valentine and Williams along with Huw Roberts Jones, founder of Byddin Ymreolwyr Cymru (The Welsh Home Rule Army) agreed to join forces and form Plaid Cymru. They took the decision following an initial meeting at a local café called Maes Gwyn, close to where I'd started my walk this morning.

The Welsh language has been undermined and under threat ever since Henry VIII passed the first Act of Union in 1536, prohibiting the use of Welsh in public administration and the legal system. In the mid-nineteenth century, the Reports of the Commissioners of Enquiry into the State of Education in Wales blamed the poor state of literacy and education in Wales on the continued use of the language in rural areas. While speaking Welsh might not have been outrightly penalised in schools, it was frowned upon and portrayed as inferior. Often this sentiment was reinforced at home by parents who felt their children wouldn't succeed in the rapidly changing modern world unless they prioritised the increasingly influential English language. Mass immigration and migration from rural Wales into newly industrialised south Wales also bolstered the use of English at the expense of the Welsh language.

By the turn of the twentieth century, concern over the erosion of the Welsh language had grown into a greater belief that Wales needed to demonstrate its own national political identity. The founders of Plaid Cymru were passionate defenders of the Welsh language and of political self-determination. The fire at Penyberth made all of Wales sit up and pay attention to their cause. As the Welsh scholar Dafydd Glyn Jones put it, this was 'the first time in five centuries that Wales had struck back at England with a measure of violence… To the Welsh people, who had long ceased to believe that they had it in them, it was a profound shock.'

I walk to the end of the promenade and then onto West End Parade, after which the Coast Path continues along an elevated track running between the dunes and the local golf club. The beach below me is empty and the sand bright white, though the footprints on it suggest a small army of early morning dog walkers have already been and gone. Marram grass swishes and ruffles as the wind flows through it.

Further down the beach, the stones are larger and flatter. I select ten of the smoothest ones I can find. The water is smooth so, for the first time in many years, I start skimming stones – varying the angle of my arm and the release of my wrist until they are skipping a satisfying distance across the surface of the water.

This simple act of play, a diversion in the middle of a journey, feels both rewarding and emotionally moving. It takes me back to a time when my kids were young and we'd spend time at the beach perfecting how to skim. Not for the first time on this journey, I feel I could just stop and lose myself for a day on this beach, relaxing into the rhythms of the sea.

This area of Pwllheli's beachfront was built over the course of just a few years to fulfil the vision of Solomon Andrews, the same man who first developed the coast at Fairbourne.

The Cardiff-based developer and entrepreneur was already a millionaire, having successfully launched new building and transportation projects throughout south Wales, including Cardiff Central Market and grand residential buildings in nearby Penarth. One summer, Andrews took a trip to the new Victorian beach resort of Llandudno. There, he heard of land for sale to the south of Pwllheli. It was marshland, of course, situated between the mouth of the Afon Rhyd-hir and the sea.

Starting in 1893 Andrews built the beachfront properties, a public bandstand, the West End Hotel and the golf course. He also constructed a horse-drawn tramway that ran along the side of the coast so that quarry rock could be transported to help shore up the sea-wall promenade. Today, Andrew's developments remain the heart of Pwllheli's south beach.

I follow the old tramway route along the dunes until I reach Traeth Llanbedrog, one of the bays regularly touted as 'prettiest in Wales' and a magnet for visitors to the Llŷn. It sits in a shallow crescent bay and is flanked, at the top of the sandy beach, by a row of multicoloured huts.

In front of me lies Plas Glyn y Weddw, the mansion that Andrews purchased in 1896 and then linked to Pwllheli via his tramway. Soon afterwards, he opened an art gallery in the house – the first of its kind in Wales. Tourists from Pwllheli flocked here to see this new, genteel cultural attraction.

In October 1927, however, the same devastating storm that destroyed parts of Criccieth also slammed into the Pwllheli coast driving the sea inland over half a mile and sweeping away Andrews's tramway. It never re-opened and visitors to Plas Glyn y Weddw dwindled.

Today there's still an art gallery in the old house but most of the visitors seem content just to wander the ornate gardens. An older couple push their grandchild around in a stroller. Ice cream melts down the child's T-shirt and falls to the ground – much to the delight of the family dog which cleans up whatever drips its way.

At the top of Mynydd Tir y Cwmwd the Coast Path emerges out of the woods and onto open ground. A brown-and-white speckled stonechat hops around the purple and yellow flowering gorse bushes, hunting for food. Abersoch lies directly ahead. In all the days of walking the Welsh coast, I've never seen so many speedboats, pleasure craft and jet skis carving up the sea. It looks more like a scene from the French Riviera than the coast of north Wales.

Seascape

Down on the beach, the tide is retreating leaving intricate foam patterns on the sand. I remove my walking boots, sling them over my shoulder and walk through the gentle surf. The water is cold but refreshing after hours of walking.

The top of the beach is lined with single-storey lodges – many featuring pristine decks and glass wind protection. Some have wooden steps leading down to the beach. Behind them sit the even more exclusive homes. It's an idyllic scene if you have the money to afford these properties. Most people who grow up in Abersoch don't.

The second home debate in this part of Wales isn't just a question of property and money. It's about community, culture, identity, political accountability and the decline of the Welsh language. With some local families living in caravans at the edge of town, and many more departed, the local Welsh-language school has been forced to close. The Welsh government recently introduced a tax on second homes with the proceeds directed into helping local communities. But it's unclear whether that will make housing affordable or even desirable for local people when so much of their community has already been hollowed out in recent years.

The second home/overtourism crisis is hardly a purely Abersochian problem. Other Welsh coastal towns like Aberdyfi, St Davids and Tenby all struggle with similar pressures (though not always to do with the Welsh language) as do coastal destinations (including some major cities) throughout the rest of the UK and Europe.

The reality is that, ever since the early days of sea-bathing tourism, the beach has attracted people and money that have transformed coastal communities. Beachfront properties continue to command the highest prices – even as increasing climate-related extreme weather demonstrates their vulnerabilities.

Today, Abersoch might be considered the Gold Coast of the Llŷn Peninsula – but for how long? Climate change sea level projections suggest that many of the beachfront properties (including all the lodges) will be at risk of storm surge flooding by 2050. It's true that the local authorities might determine the real estate is too valuable not to protect but how appealing will the beach be once the necessary coastal defences have been built? And will the owners even be able to insure those fancy holiday homes in the future? Plenty of beachfront residents in America would already say no.

The United States now averages a $1 billion natural disaster every eighteen days and climate change is displacing millions, not just because of catastrophic events but for the mundane and very real consequences of living in coastal areas where insurance companies no longer offer coverage. Six Florida insurance companies went bankrupt in 2023 and home insurance premiums in the state rose over 300 per cent from 2018 to 2023. Some industry experts now believe that Florida's coastline will be uninsurable in just a few years. This situation will be similar the world over, including here in Wales where thousands of properties are at risk.

I have one more trip along the Llŷn Peninsula to accomplish – a thirteen-mile walk from Porthdinllaen, where William Madocks envisaged his ferry crossing to Dublin, to the village of Trefor, just south of Caernarfon. I've mapped it and, frankly, it looks like a serious hike not least because of one almighty climb from sea level up and around Garn Canol.

I set off from Morfa Nefyn beach. To my left, I can see the pretty bay of Porthdinllaen and the red façade of the Tŷ Coch pub, nestled among fishermen's cottages at the base of the curving promontory.

This sheltered bay once was the hub of the Llŷn's salted herring industry, much of which was then processed in nearby Nefyn. Nowadays the main catch is tourists who come to Instagram themselves with the pretty bay and the red pub as a backdrop. On the surface then, Porthdinllaen's marine importance might appear to have passed. Under the surface though, on the shallow seabed to be exact, this bay may yet prove to be an essential asset in helping Welsh coastal communities mitigate the impact of climate change.

Porthdinllaen bay is considered so important because it boasts a large seagrass meadow. Like the intertidal salt marsh, these plants act as carbon sinks – they absorb carbon dioxide into their leaves and lock it in as organic matter through photosynthesis. They then pass carbon into the marine bed that they grow in. Seagrass meadows also help protect fragile coastlines from storm surge erosion, help support entire ecosystems of marine species and provide a nursery ground for fish.

Seagrasses are found along the shallow coasts of temperate and tropical seas all over the world and cover up to 372,000 square miles. That's only 0.1 per cent of the world's seafloor but it is responsible for up to eighteen per cent of the organic carbon buried in the ocean. Taken as a whole, they produce more oxygen than all terrestrial rainforests and grasslands combined. However, as with so much of the natural world in our modern age, the global seagrass coverage is threatened by marine pollution. Some ninety per cent of UK seagrass has been lost during the last century including large tracts along the Welsh coast.

Over the last couple of years, scientists have started harvesting seeds from the Porthdinllaen seagrass and replanting them in depleted Welsh marine areas. The hope is that, by borrowing from Porthdinllaen, Wales may once again benefit from the protection this sea flora provides.

I leave the bay and make my way towards Nefyn. The cliff-face erosion is severe in parts – so much so that the path is diverted inland. When I stop for a coffee at the village's beachfront café, the young woman in charge asks if I will sign a petition, started by local people, to have the eroded Coast Path rebuilt by the Welsh government.

'It's so important for the tourist trade. It will be a disaster if they don't repair it,' she says.

The twin towns of Nefyn and Morfa Nefyn have recently supplanted Abersoch as the places on the Llŷn with the highest number of second homes. The prices are mind-blowing. A seven-bed detached home is on the market for £1.25 million, a six-bed detached has an asking price of £995,000. Then there's the four bed, four-bathroom house for £950,000. The average salary in the area is around £20,000 a year. It's easy to see why local people feel muscled out.

Just five miles inland, house prices are nearly £100,000 cheaper than here on the sought-after coast. That is still out of the price range of many local families but it also shows how the lure of the sea, and the disparity between urban and rural incomes, is causing a crisis for local communities and the cultural heritage of the Llŷn.

At least one dwelling in Nefyn clearly isn't a second home. It boasts a large banner hanging in front of the house that reads: *Yma o Hyd* (We are still here).

This is the title of a defiant rallying song written by the Welsh folk singer and nationalist Dafydd Iwan. He wrote it back in 1983 at a time when Wales was in the stranglehold of an economic recession and the policies of the English Prime Minister, Margaret Thatcher, were destroying the livelihoods of Welsh coal miners and steelworkers. It also came on the back of defeat for a referendum that would have given Wales devolved political powers from Westminster. And it was written during a time when clandestine activists and agitators had started fighting back against what they saw as the anglicised gentrification and linguistic ruination of rural Wales. Their weapon was arson attacks on English-owned second homes, echoing the Tân yn Llŷn protest of the 1930s. The very first attack by a group calling itself Meibion Glyndŵr (the Sons of Glyndŵr) took place just north of here at Tyddyn Gwer in Nefyn in 1979.

Within Welsh language and Plaid Cymru circles, '*Yma o Hyd*' quickly became a totemic anthem and rallying cry. Its lyrics hark back to the legacy of Macsen Wledig (Magnus Maximus) the Emperor of the West and the last Roman ruler of Britain. Macsen is legendary in Welsh history because it is claimed that, when his army withdrew from the lands of Wales, he transferred power back to the local people. By the time that the oral history and mythic stories of the *Mabinogi* were being written down, Macsen's reputation had grown considerably. He was said to have married a beautiful local woman named Elen and, through their offspring, started a Welsh royal lineage.

'*Yma o Hyd*' became popular throughout the Welsh-speaking and Welsh nationalist community but it hardly resonated with English-speaking Wales though it had grassroots popularity with various local football and rugby fans. Then, seemingly out of nowhere, Welsh football fans started singing '*Yma o Hyd*' at international games. '*Yma o Hyd*' became a football anthem during Wales's successful campaign to qualify for the 2022 World Cup and Dafydd Iwan became a star in his own right. Today, forty years after '*Yma o Hyd*' was first sung by Dafydd Iwan, it is one of the best-known Welsh-language songs in Wales.

I pass more holiday cottages as I climb out of Nefyn towards the next village, Pistyll. The path emerges out into wide, open hillside that slopes gently down towards the cliff edge.

At St Beuno's Church (named for another seventh-century Celtic saint who is known as the patron saint of sick cattle apparently), the path heads out over the headland, providing a view of open water stretching all the way to Holyhead at the tip of Ynys Môn. In the bay, hundreds of gulls swoop close to the water, their interest sparked by some intense activity under the surface. Two brown cows graze by the cliff edge, watching the kerfuffle below with placid indifference. Clearly, they've seen it all before.

Ahead of me sits a long, gentle crescent bay. At the top of its pebble beach lies a set of old, alien-looking buildings that sit below a hollowed-out cliff face. The ghosts of old industry litter the landscape wherever you walk in Wales. But this is different – it's an excavation on a scale I haven't before encountered along the coast.

Only after I climb the path away from the beach do I realise I'm at Nant Gwrtheyrn, once one of the most important quarries in all of Europe. This bay was the source of the granite that paved the streets of cities throughout the Victorian world. By 1886 more than 200 people lived in this isolated coastal village – despite the fact that land access was restricted to a single track that snaked up over the mountain, navigable only by foot or a very fit horse and cart.

The quarrymen of Nant Gwrtheyrn had particular skills and tasks. There were the Rock Men who drilled the holes for the explosives to blast out the granite. The Pop Drillers and Breakers used hammers and gunpowder to break up the giant slabs. The Blockers determined how many sections of granite were formed in a slab and the Settsmen (the artisans of the process) used dressing hammers to shape the brick-like setts that were shipped to the new cities.

Despite their isolation from the rest of the Llŷn Peninsula, the locals were connected to the outside world through the ships that came to collect the granite. These brought provisions from cities like Liverpool including linen and fine dresses for the women of Nant. In fact, it was said that the ladies of Nant were the most glamorous of all the women on the Llŷn Peninsula.

The granite industry fell into decline after the First World War and most of the families started to leave Nant Gwrtheyrn. The

Seascape

last sett was cut around the outbreak of the Second World War and the last family left this place of splendid isolation by the late 1950s. For a short time in the 1970s, the abandoned village was taken over by a London hippy commune that called itself the New Atlantis.

When the hippies departed for another site in Ireland, Nant Gwrtheyrn would probably have become a ruin were it not for the vision of one man. In 1970, an English-born physician named Carl Clowes moved from his home in Manchester to Llanaelhaearn, a small village a few miles inland of here. Clowes's mother was Welsh and, once he'd moved to the Llŷn, he and his wife were keen to raise their own children as Welsh speakers. The first Welsh Language Act had come into effect in 1967, a legacy of the Welsh nationalist agitation and campaigning that began here on the Llŷn. This meant there was a growing need for Welsh-speaking staff at public organisations across Wales – including hospitals and doctors' surgeries.

Clowes realised there was an opportunity to teach Welsh to adults and so, for the next eight years, he spearheaded a community-led campaign to raise the £25,000 needed to purchase the derelict hamlet of Nant Gwrtheyrn. People from all over Wales contributed and Clowes's ambition and persistence galvanised a growing national consciousness of the importance of preserving and regrowing the Welsh language. In 2003, the Nant Gwrtheyrn National Welsh Language and Heritage Centre opened its doors to the public, offering immersive courses for adults wanting to learn Welsh. The day was marked by a concert featuring the actor Rhys Ifans and the acclaimed Welsh indie rock band Super Furry Animals. Two of their members, Cian Ciaran and Dafydd Ieuan, had a special reason for performing. Carl Clowes was their dad.

It's time to move on towards Trefor. That means climbing out of Nant Gwrtheyrn up what I can already see is a stupidly steep road incorporating multiple switchbacks.

A car pulls up next to me as I start walking. The driver, a female priest, rolls down her window and says: 'I can give you a lift up the hill if you like.' Before I even have time to answer she continues, 'But you look like you want the challenge of conquering it on your own.' I'm really not sure what I want, to be honest, but now I feel I have no choice.

'Yes, it's all part of my day's walk,' I reply without much conviction.

The priest nods approvingly and drives off up the hill.

This is probably the warmest day of the year so far and I'm feeling all of my fifty-seven years as I trudge up the valley. Only later do I learn the history of this road – how it wasn't until the 1930s that a motor car succeeded in driving up the hill, so steep and severe was the gradient in some places and so sharp the switchbacks.

The last two miles are excruciating. The inside of my legs are chafing and I take each step with gritted teeth. At the highest point of this section of the path, between the peaks of Garn Fôr and Garn Ganol, I meet a couple walking with their dog. They warn me that the route down into Trefor is closed off because the *Game of Thrones* prequel, *House of the Dragon*, is being filmed there.

I'm in no mood at this point to turn back. I'm going to take the shortest, most direct route down into Trefor even if there be dragons. Luckily the film crew have departed by the time I start my descent so I shuffle down the hill without incident. I head to Y Tafarn, the lone pub in town. A flock of bedraggled sheep mooch around in the front garden much like those teenagers outside St Davids Cathedral, while the landlord clears up the debris from a raucous birthday party the night before.

While I wait for him to open, I have time to contemplate the language and demographic changes that are taking place here on Llŷn and in other parts of north Wales. For centuries, the peninsula's isolation from the rest of Wales along with its connected coastal towns and villages helped preserve both the Welsh language and a strong sense of community. Back in the 1920s, it made perfect sense for the founders of Plaid Cymru to choose the Llŷn to show their act of defiance and send out a rallying call for the defence of the Welsh language.

Wales has changed a great deal, though, over the past 100 years. There has been continued demographic flight away from rural and coastal communities to the few large cities and out of the country to other parts of the UK and beyond. The challenge to preserve and grow the Welsh language won't be won just by shoring up local communities here on the Llŷn. It needs a whole country initiative.

Political devolution and the formalisation of Welsh as the official language of government laid the foundations. Widespread investment in Welsh medium schooling throughout Wales has also been an important and impressive achievement – exposing a new generation of young people to speaking Welsh every day within school. The ultimate challenge, though, remains to integrate Welsh into the fabric of everyday life in Wales.

Perhaps the fluid demographics of our climate future might offer a chance to further restore the Welsh language. Already many new immigrants to Wales are committing their children to Welsh medium schooling. For these new arrivals, it's a functioning language of Wales just as much as English is. As climate change sends more refugees to our shores, adding once more to the melting pot of Wales, an opportunity exists to grow the Welsh language in parts of the country where English language migration decimated it 200 years ago.

Y Tafarn has opened and so I chat with the landlord about all the filming taking place around Trefor as he pours me a pint. I ask him if any of the actors have come in for a drink.

'Oh yeah, we had Rhys Ifans last week. He's filming that

House of Dragons show and he popped in for a couple,' says the landlord, before adding: 'Personally, I'm not that fussed with all these celebrities. As Rhys said, "You're the only pub that doesn't want my photo on the wall."'

12

Two Bridges over Troubled Waters

Walks from Caernarfon to Menai Bridge and Niwbwrch to Aberffraw

In the last years of the thirteenth century, an architect from the County of Savoy, part of the Holy Roman Empire, arrived in north Wales with a very special brief. His name was Jacques de Saint-Georges d'Espéranche (referred to as James of St George or James the Master Builder in English). He had been commissioned by the English King, Edward I, to build a set of castles that would subdue the Welsh resistance in Gwynedd and demonstrate the overwhelming might of the English crown.

Edward had defeated Llywelyn ap Gruffudd (Llywelyn the Last) in 1282. It had taken over 200 years to fully suppress Welsh resistance but now all of Wales was under English rule for the first time. James's reputation as a military architect was the talk of Europe, and Edward had first met him when he returned through Savoy from fighting in the Crusades.

James would design and oversee the construction of seven castles in north Wales including Harlech (with its steps leading up from the sea), Conwy, Beaumaris and Caernarfon. His military architecture was renowned for its intricate design – including inner and outer protective walls as well as circular towers – which made the castles he built almost impregnable to attack.

James's inspiration was the bastide towns of Gascony in southwest France where the town and its activities were integrated into the castle. Here in north Wales, the new English settlers lived within the protection of the castle walls. The local population could enter by day but could also be locked out at night.

I start today's walk outside Caernarfon Castle. My plan is to meander up the Menai Strait to the two bridges that connect the Welsh mainland to the island of Ynys Môn (Anglesey).

Even today, some 700 years after it was built, Caernarfon Castle remains formidable.

The castle was conceived not just as a fortress but also an elegant demonstration of authority. It featured eagle statues, polygonal towers and multicoloured masonry. Master James of St George also erected tall stone town walls and created a quay by the side of the castle on the Afon Seiont – partly to bring in the vast amounts of stone needed. The mega project took forty-seven years to complete and costs spiralled to £25,000 – about £17 million in today's money.

Nowadays, we're used to hearing about budget overruns on major infrastructure projects like the HS2 and Elizabeth Line railway networks. It seems that things weren't so different back in Norman times. All told, Edward I's Welsh castles cost the crown almost £100,000 to build – well over twice the king's annual income.

Once completed, Caernarfon Castle served as Edward's Welsh seat of power and administrative centre and sent a not-so-subtle hint to his new subjects.

'Caernarfon stands out, too, in its echoes of Rome and Constantinople – deliberate expressions of its imperial role,' wrote John Davies in *The Making of Wales*.

Seascape

In evoking Roman times, Edward may also have been reminding the Welsh of Macsen Wledig who dreamed of a great fort at the mouth of a river – 'the fairest that man ever saw' – where he fell in love with Elen. She, at this point, had been embraced by the Welsh as an important Celtic saint.

As with its sibling fortress, Harlech to the south, Caernarfon Castle's coastal location was strategically important. Not because Edward feared attacks from the sea. The days when the Welsh would raise a fleet to take on the Normans (as happened in 1090 further up the Menai Strait when Gruffudd ap Cynan, with the support of a Viking king, Magnus 'Barefoot', defeated the Earl of Chester and the Earl of Shrewsbury) were over.

Instead, access to the sea allowed the castle towns to be resupplied and their Anglo-Norman inhabitants to travel without risking their lives by going overland. Caernarfon also allowed easy access across the mouth of the Menai Strait to the agriculturally rich but geographically isolated lands of Ynys Môn, the island referred to by Gerald of Wales as Mam Cymru – Mother Wales. Caernarfon soon became one of the most important ports in early medieval Britain.

The quay is still in use. Years ago I brought my kids here to tour and learn about the castle. To be honest, they had more fun trying to catch crabs off the edge of the dock. I spent the afternoon hovering behind the two of them as they sat, feet dangling over the side, ready to make my own catch if one of them fidgeted too much or peered a little too far down into the water some ten metres below.

I spend a few minutes walking around the quayside, soaking up these memories. There are a couple of small boats tied up and a smattering of dog walkers taking in the morning air. There is a peaceful, retired feel to the place – completely different to what life must have been like nearly two centuries ago when this was a vibrant working dock exporting slate around the world.

In the 1850s, some 250 ships were involved in the export of slate from Caernarfon, and nearly 3,000 vessels visited the port every year. The docks boasted many ancillary industries including ship building, rope making, ship broking, joinery and carpentry. At peak sailing times, some 10,000 people lived in the old town and it was a rough and tumble place. It teemed with cramped temporary lodgings, pubs and brothels.

This would have been an intimidating place for any newcomer to adjust to – even more so if you were a young woman. Luckily for twenty-year-old Ellen Francis, she had grown up in Amlwch, a tough port on the north coast of Ynys Môn. Francis arrived in Caernarfon with a unique pedigree. Her father, Captain William Francis, was a renowned mariner who had set up a school in Amlwch to train local men how to navigate at sea. He also trained his children, including Ellen.

Soon after Francis arrived in Caernarfon, sometime in the 1830s, she opened her own maritime navigation school. She married soon afterwards and became Ellen Edwards. Her reputation grew rapidly and soon her maritime school was flourishing. She was a devout non-conformist – a faith that would have inured her against the debauched excesses of port life but also put her at odds with the Anglican Church of England assessors, who gave her a damning review in the infamous Reports of the Commissioners of Enquiry into the State of Education in Wales.

Despite these slights, Ellen Edward's nautical academy flourished and her reputation spread throughout north Wales and beyond. Her value to Caernarfon was firmly cemented with the passing of the Mercantile Marine Act of 1850. Any officer on board UK vessels sailing abroad had to pass the examination, and Edwards, along with Sarah Jane Rees from Llangrannog (better known by her bardic name, Cranogwen) were two of the most important female instructors. It is estimated that Edwards trained more than 1,000 mariners over sixty years of running the academy. She also taught her own daughter those same navigation skills and she took over the running of the academy when her mother retired.

The contributions of both Ellen Edwards and Sarah Jane Rees are just the most high-profile examples of the way women were integral to Wales's port towns. They tended to be far more involved in commercial activities than their counterparts in inland and upland rural villages. Often, their husbands and fathers were away at sea and it was left to the wives and daughters to not just hold the family unit together but to help with the daily running of the community and with its religious and spiritual direction.

I leave the old docks and walk around the outside of the castle walls, passing the stone arch leading to High Street and heading north up the Menai Strait. Back in Ellen Edwards's day, this waterway would have been crammed with international ships. Despite the squalor in the old town, it must also have been an exciting, intoxicating place to be a part of.

A while back, I had listened to a sea shanty called '*Llongau Caernarfon*' ('Ships of Caernarfon'). It captures the essence of nineteenth-century life in this port town and tells the story of a young boy, watching with his mother as the ships leave the harbour.

Translated into English, the first lines are:

> All the boats by the quay are loading
> Why can't I go like everyone sailing.
> There are three starting to raise the anchor
> And sailing tonight for
> Birkenhead, Bordeaux and Wicklow.

'*Llongau Caernarfon*' is an evocative song about the town's heyday but it wasn't written in the nineteenth century. It was penned by the poet and scholar J. Glyn Davies in 1934 as part of a collection of reimagined sea shanties and folk songs.

Wales has a strong folk music tradition but not many songs are devoted to the sea. *'Ar Lan yr Môr'* ('On the Seashore') is perhaps one of the best-known ones but even that was only first recorded in 1937. Another, older, maritime song is 'The Bardsey Boat Lament', written after a local boat called *Supply* sank near Bardsey Island, off the Llŷn Peninsula, in 1822. A 1906 recording of a local man singing the song is kept in the National Museum of History at St Fagans. 'The Bells of Aberdovey', an ode to Cantre'r Gwaelod, is older still. The melody was composed back in 1785 and the lyrics were written in the early nineteenth century.

Other examples of Welsh maritime songs are harder to find – probably because they were passed down in an oral tradition by sailors and fishermen rather than written down. As that way of life faded, so did the memory of these songs.

Davies's songs were new creations but they preserved and evoked the essence of life at sea and around the Welsh coast. Many of the songs in his two collections of sea shanties, *Cerddi Huw Puw* and *Cerddi Portinllaen*, drew on melodies he heard Caernarfonshire sailors sing when they drank at the Welsh Harp tavern in Liverpool at the turn of the twentieth century, and also from his experience working for the Cambrian Shipping Company.

Davies observed that Welsh sea captains operating out of Liverpool and north Wales's ports would carefully recruit young men who had grown up together, had sung in the choir and for whom the sense of three-part harmony was natural.

Today J. Glyn Davies's sea shanties are being played by a new generation of folk musicians keen to remember and celebrate Wales's maritime past – one whose stories might have been lost were it not for his imagination.

I listen to some of Davies's songs as I walk up the Menai Strait, including the very whistleable *'Fflat Huw Puw'*. It puts a spring in my step as I follow the Coast Path through the old slate port of Y Felinheli before arriving at Britannia Bridge, built by the famed civil engineer Robert Stephenson in 1850 and one of two nineteenth-century engineering marvels that connect Ynys Môn with the mainland.

Ahead of me I can see the other crossing – the Menai Suspension Bridge – completed twenty-four years earlier by the equally famous engineer Thomas Telford. It was part of his grand infrastructure dream of building a road network across north Wales to Holyhead on Ynys Môn. Between the two bridges lies the Swellies – reputedly one of the most treacherous stretches of water in all of Wales.

Some 14,000 years ago, it would have been possible to walk across the Strait, but melting glaciers and the subsequent rises in sea level gradually flooded the dry land until, around 5,000 years ago, the waters fully cut off Ynys Môn from the mainland. Ever since, the Strait has been influenced by a double and reversable tidal flow that can cause havoc for anyone trying to navigate the part of the channel where the two tides meet. Simply put, the tide flows up towards the southern mouth of the Strait near Caernarfon and from there it heads northward. At the same time, the tidal wave continues to flow past the western coast of Ynys Môn (it moves quicker in open water) and enters the Strait from its northern end near Bangor. The result is that two parts of the same tide flow towards each other until they meet, creating a clash that generates strong currents and whirlpools.

The unpredictable nature of the Swellies is what made travel from Ynys Môn to the Welsh mainland so dangerous. Sometimes you could almost wade across the Strait. Other times, the strong currents would sink boats; in 1785, a ferry carrying fifty-five people ran aground on a sandbank. Only one person survived.

Telford faced opposition from local sailors and fishermen who feared a bridge would obstruct the safe passage of shipping. So he designed a suspension bridge – the biggest in the world at the time – featuring sixteen chains that held up 579 feet (176 metres) of deck. This allowed 100 feet (30.5 metres) of clear space beneath for ships to pass under. It also helped reduce the journey time between London and Holyhead from thirty-six hours to twenty-seven.

Stephenson's Britannia Bridge met a different objective – connecting Holyhead to the fast-growing rail network being laid down across the British Isles. It too had to be tall enough to enable ocean-going ships to pass under so Stephenson, along with a fellow civil engineer, William Fairbairn, experimented with the concept of riveting together iron tubes to carry the necessary weight and cover the span of the Strait. It was a pioneering technology that Stephenson would later deploy in countless other bridge-building projects.

Telford and Stephenson's achievements at the Menai Strait would cement their reputations as some of the Victorian Age's greatest engineers. When they died, they were buried next to each other in Westminster Abbey. Yet even the most ingenious innovations are sometimes no match for the unpredictable force of nature known as bored teenage boys. In 1970, a group of them decided to explore the tubes underneath Stephenson's railway bridge. It was dark so one of the boys used a lighter to set fire to some paper. Within hours, the entire bridge was on fire and structurally compromised. It would take ten years before repairs were complete and the bridge could re-open. I imagine Stephenson (and Telford for that matter) would be turning in his grave at the thought.

In 1561 Queen Elizabeth I issued an injunction that permitted the mayor and bailiffs of Niwbwrch (Newborough), a small community on the southwestern tip of Ynys Môn, to punish whoever was found cutting, uprooting or carrying away the marram grass that grew in the extensive sand dunes on the edge of the town. Furthermore, her government ordered the planting of new marram to shore up the dunes.

Seascape

For Elizabeth's government, and the residents of Niwbwrch, it was a direct way to mitigate the effects of the major storm systems that battered the Welsh coast during this time and anchor the sand that threatened to swallow up the hamlet. In our modern, climate change impacted world, Elizabeth's act would be classified as a prime example of environmental protection, employing what sustainability experts call 'nature-based solutions'.

I've come to Niwbwrch to explore these dunes and walk the long crescent beach of Traeth Llanddwyn before heading west towards the village of Aberffraw, once the capital of the Kingdom of Gwynedd and the last stronghold of Llywelyn the Last.

The glistening sun casts in silhouette the families playing on the beach down by the water's edge. Behind them, across the water, the hills of the Llŷn rise up in the distance. I grimace a little at the thought of having walked over them a couple of days before.

At the far end of the beach is the islet of Ynys Llanddwyn, named after Dwynwen, another of those Dark Ages Welsh saints who sought out a hermitage by the coast – her isolation borne out of a desire to escape a forbidden love. She dedicated her life to worshipping God on this rock looking out to sea. Today, she is considered the Welsh patron saint of lovers – which has always struck me as a little odd given her dedication to abstinence.

A little further down the beach, a brother and sister – no more than eight or nine years old – pull themselves up one of the steeper dunes using an old piece of rope slung around a pine tree. The sounds of people playing catch and walking their dogs are carried up over the dunes by the wind.

When these dunes were created during the storm cycles of the thirteenth and fourteenth centuries, the hamlet of Niwbwrch was just taking shape. It had been founded in 1294 by former residents of Llanfaes who had been evicted by Edward I so he could build Beaumaris Castle on their land. Previously, it had been a Llys (a royal court and administrative centre) for the Princes of Gwynedd known as Rhosyr. Back in the 1980s, archaeologists unearthed the ruins of the Llys, buried under sand on the outskirts of the modern village.

The new inhabitants were attracted to Niwbwrch by its fertile agricultural lands but storms threatened its productivity. Historical records cited in a 1960 *Journal of Ecology* paper describe how: 'About one third of the land of [a local] manor was damaged by storm on the feast of St. Nicholas, 6th December 1331, when 186 acres were destroyed so thoroughly by the sea and inflow of sand as to render it useless for agriculture evermore.'

Many families were driven out of their crofts from land between Llanddwyn and Niwbwrch and it was during one of these storms that Llys Rhosyr was lost to the dunes.

Marram grass was planted to stabilise and slow the encroaching sands. Soon, rabbits started colonising the dunes – providing a reliable form of protein for local villagers and also creating warrens that allowed invertebrates, lizards and snakes to thrive. It didn't take long for local people to realise that marram, when dried, could be weaved to create ropes, nets, mats and baskets – which they did to such a scale that they steadily degraded the dunes. This was what prompted Queen Elizabeth's environmental intervention.

Over time, marram harvesting started up again and the structural integrity of the dunes was again compromised. Following the end of the Second World War, the UK Forestry Commission planted Corsican pine trees thinking they needed to anchor and stabilise the dune system and to meet the post-war demand for timber. As often happens, though, when a non-native species is introduced, the growth of the forest had unintended consequences. Some of the marram grass perished because the pine trees crowded out the available light. The trees also sucked up large quantities of fresh water – lowering the water table and further undermining the dunes. On the positive side, the forest became a haven for the endangered red squirrel.

Fast forward to today and, despite the best efforts of the Forestry Commission, the fragile sand dune ecosystem still boasts a wealth of biodiversity. I walk along the beach following the dunescape. Sea spurge, sand cat's-tail and dune pansies sprout up amid the marram grass. Further inland orchids, butterwort, grass-of-parnassus and yellow bird's-nest prosper.

As extreme weather intensifies, the shifting sands could pose serious problems for the people of Niwbwrch. Here, in the dunes, another part of the Sands of Life preservation and coastal protection project that I'd seen down in Harlech hopes to shore up the seashore for at least the near future.

The goal isn't that different from what it was in Elizabethan times: allow the dunes to breathe and move naturally so that they remain contained in their own ecosystem and not spread through the surrounding area. The Sands of Life work here involves removing invasive scrub, fallen trees and old stumps from areas within the forest. Doing so will help wildlife to thrive, including the rabbit population which has steadily been rebuilding after its near eradication in the 1950s from the highly contagious viral disease myxomatosis. Protecting the integrity of these dunes isn't just about stopping the sand swallowing Niwbwrch. It is about preserving an entire ecosystem.

The long-term future for Niwbwrch beach and its surrounding area remains uncertain, however. The sands are in constant flux – in one six-week period in 2020, the sea advanced inland by six metres. With no commercial or residential properties directly at risk, it's unlikely that the authorities will want to invest in a future fight against nature.

At the far perimeter of Niwbwrch forest, the Coast Path joins a long causeway spanning the Afon Cefni Estuary called Malltraeth Cob. Malltraeth roughly translates as the blasted beach. In fact, it's less a beach and more a vast salt marsh. Wild ponies graze on the inland side of the Cob, their manes swaying gently in the breeze.

Two Bridges over Troubled Waters

The Cob, similar in design to William Madocks's creation down in Porthmadog, was first proposed in the 1750s to help reclaim land for agricultural use. Work started immediately on the causeway and the canalisation of the Afon Cefni. You probably won't be surprised to learn that the project ran into difficulties. A major storm on the night of 17 October 1789 damaged the sea defences and work was halted. It might never have been built were it not for the fact that Thomas Telford had planned his grand A5 trunk road to pass this way. He designed tidal sluice gates (similar to those created by the clever Norman monks down on the Gwent Levels) which allowed some 4,000 acres of land to be reclaimed from the sea.

Despite Telford's engineering feat, local people remained all too aware of the Malltraeth Estuary's history. A folk song, '*Cob Malltraeth*', was popular on the island in the nineteenth century. An English language translation of the final stanza goes like this:

> If Malltraeth cob breaks, my mother will drown;
> I fear it in my heart ti-rai, twli wli
> I fear in my heart that I shall be the one to suffer.
> I can neither patch nor wash my shirt;
> I fear it in my heart, ti-rai, twli wli wli ei,
> I fear in my heart that I shall soon perish.
> But, thank heaven, the old lady was seen
> Safely taking refuge, ti-rai, twli wli wli ei,
> Safely taking refuge in the shelter of the rock.

I pause on the Cob and look ahead to where construction workers are shoring up the old stone wall sea defence. Based on some 2050 projections, which show flooding reaching as far inland as Llangefni, five miles away, I'd say they have their work cut out for themselves.

I've seen many examples of how the sea has reshaped the coast during my walks around Wales but the dune and marsh complex on either flank of Aberffraw takes a while to get your head around. It's a continuation of the Niwbwrch dune system and it stretches nearly two miles inland. Here, the Coast Path offers two routes to the village. At low tide you can walk along Traeth Mawr – a stunning wide sandy beach (no great surprise given how it was created). At high tide, the smart option is to cut directly through the dunes until you reach the banks of the Afon Ffraw.

In the late thirteenth century, this was the most important place in all of Wales. It was the seat of power for the House of Aberffraw and the Princes of Gwynedd, Wales's dominant royal family. The dynasty had been established back in the ninth century when Rhodri Mawr (Rhodri ap Merfyn) inherited and amalgamated the kingdoms of Gwynedd, Powys and Seisyllwg. Rhodri then divided these combined kingdoms into provinces to be ruled by his sons on his death. Rhodri received the Kingdom of Powys in mid-Wales. Cadell ap Rhodri got Deheubarth and created the House of Dinefwr based at Llandeilo in south Wales. Anarawd ap Rhodri gained the Kingdom of Gwynedd and founded the House of Aberffraw.

It was from here that Anarawd's descendant, Llywelyn ap Iorwerth (Llywelyn the Great), consolidated power to become the dominant Prince of Wales in the thirteenth century. But Aberffraw's powerbase was destroyed following the defeat of his grandson, Llywelyn ap Gruffudd (Llywelyn the Last), by Edward I in 1282. With the Gwynedd dynasty defeated, the royal court fell into disarray. A final insult came in 1316 when, depending on who you talk to, the buildings were dismantled to provide stone for the construction of either Caernarfon or Beaumaris Castles. There was almost nothing left of the royal court by the time the storms reshaped this coast. But I can't help but think that the House of Aberffraw would have preferred to be swallowed by the sand than become part of Master Builder James of St George's master plan.

Two Bridges over Troubled Waters

The Normans understood the importance of taming Ynys Môn. No wonder they built a formidable castle at Beaumaris – a strategically crucial part of the Menai Strait. That turbulent waterway had long set Ynys Môn apart from the rest of Wales and given it a sense of power. The House of Aberffraw had embodied this aura – part of a legacy that went back to ancient times when the revered Druids were said to have maintained their seat of mystical power among the oak groves of the island. To defeat them and to crush their spirits, the Romans sacked Ynys Môn and destroyed those oak groves.

The bridges built by Telford and Stephenson nullified the menace and barrier of the Menai Strait, establishing easy access to the island and bringing with it waves of cultural and demographic change. Yet, even today, when you wander this beautiful coastline on the outskirts of Aberffraw, with views far out into the Irish Sea, you can sense an air of distance, defiance even, to the mainland. It remains an island apart.

13

The Wreck that Inspired the Shipping Forecast

A walk from Amlwch to Moelfre

There is one more stretch of the Ynys Môn coast that I am keen to accomplish – the northern trail from the old copper mining town of Amlwch down to Moelfre.

It's a sleepy early autumn morning when I arrive in Amlwch and head down to the old port. A few middle-aged men are tinkering on small boats in the equally small harbour but, apart from that, all is quiet. I walk out onto the headland and meander around the craggy set of bays and nooks that make this part of the island so appealing.

Inland, a couple of miles away, sits Mynydd Parys – the main reason Amlwch and this port exist. Its nickname is Copper Mountain because, for the past 4,000 years, local people have been tearing chunks out of it to get valuable copper ore. In the late eighteenth and nineteenth centuries, it became the largest copper ore exporter in the world – much of which was sent by ship to the copperworks in the Swansea valley. Today, the mountain looks like a scorched red, gold and brown alien landscape. No wonder it's been a location for *Doctor Who*.

Amlwch had a population of over 10,000 at the peak of its copper mining boom. At that time, it was Wales's second largest town after the iron capital of Merthyr Tydfil. Michael Faraday, the great electricity pioneer, visited Parys in 1819 and recounted its potential:

'Here the vein had swelled out into a bunch ... and had afforded a very rich mass of ore. Here again it became very narrow and we had in one corner to lay down on our backs and wriggle in through [a] rough slanting opening not more than 12 or 14 inches wide. The whole mountain being above us and threatening to crush us to pieces.'

When the boom ended, Amlwch reinvented itself for a while as a shipbuilding centre but that business also faded. So, as with so many of the old industrial towns and villages that I've visited along the Welsh coast, it would be easy to assume that Amlwch's days shaping the future of global industry and enterprise are just a memory.

Except that Parys Mountain might still have an important industrial contribution to make. The old copper mine helped drive innovation in the shipping and electric telegraph industry. Fast forward nearly two centuries and copper is once again playing a crucial role, this time in the renewable energy industry – its conductive properties help power wind turbines, generators and transformers to mention just a few of its uses.

Recent geological surveys suggest copper mining could still be viable at Parys. They also identified significant deposits of zinc, a metal that is seen as a more sustainable and affordable substitute for lithium in the next generation of batteries needed to store the energy generated from renewable sources.

I stop for a break at Porth Eilian. There's a bench on the path overlooking the secluded beach. Ahead of me, the Trwyn Eilian

Seascape

lighthouse stands tall at the very edge of the headland. As I sit, I muse a little on Amlwch's potential to rise again.

I've walked through so much of Wales's industrial heritage (and decay) over the past few months, whether it's the ancient lime kilns dotted along the west Wales coast, the old coal seams that are exposed in the cliffs at Amroth, the silted-up iron and copper dock at Briton Ferry or the granite works at Nant Gwrtheyrn.

All these industries brought economic progress to Wales and great wealth to their owners. But they also brought great hardship for the people who toiled to create that wealth and, for the most part, they exacted a heavy toll on the environment and landscape of Wales.

Is it possible to undertake a new form of copper or zinc mining for the renewable energy sector that helps society without further destroying the mountain? It is hard to believe that people would be in favour of new mineral extraction if it takes such a toll on the physical environment as Mynydd Parys has suffered over the years. Especially now that Amlwch's copper story is as much a part of its new tourism trade as its extractive past.

Another industry here on Ynys Môn rivalled the copper trade. That was smuggling. In the eighteenth century, the illicit trade of contraband goods exploded after the English crown imposed high import taxes on both luxury and staple goods such as salt, brandy, wine and even soap and candles.

I'd walked through historic smuggling hotbeds on previous journeys. On the south Gower coast, I'd come across the old Salt House at the far end of Port Eynon beach. In the eighteenth century it was said to have been a fortified hideout for one of the the area's most notorious smugglers and strongmen known as John Lucas. One night, so the story goes, Lucas's gang were unloading a smuggled shipment of tobacco, brandy and French lace near the rocky headland when they saw lights on the beach coming towards them. It was customs officers – perhaps tipped off in advance. Gun shots rang out and the smugglers tried to escape with their mules loaded down with contraband. Then more shots and one of the smugglers who had been holding the boat tumbled back, dead, into the water. He had been shot in the head.

The gang escaped around the headland and the customs officers gave up chase. Locals believe they hid in a cave just around the headland known as Culver Hole (and not far from Paviland cave), where Lucas had dug a secret tunnel connecting back to the Salt House.

Cwmtydu in Ceredigion, where Jeff, Tim and I had walked, was a salt smuggling centre. Traeth Morfa Bychan, just round the bay from Porthmadog, was a particular smuggling favourite because of the series of caves at the north of the beach where contraband was stored before being fenced in the local community.

On Ynys Môn the smuggling trade was particularly rife – partly because of the large number of isolated coves and beaches (like here at Porth Eilian), partly because of the long-held animosity to the English crown (even more so than in other parts of Wales) but especially because of the island's geographical proximity to the Isle of Man, just under fifty nautical miles away.

The Isle of Man was not officially part of Britain and so provided the perfect staging post for stockpiling contraband goods waiting to be smuggled into Wales. Gangs would travel over to the island then return in the dead of night in boats laden with spirits, salt and other taxable goods.

One 1763 account by the master of a revenue cutter (fast customs ships that were built to outrun and catch smugglers) outlines how the gangs worked. He describes how they travelled independently as passengers on scheduled boats to the Isle of Man and then met up as a group before hiring Irish wherries to take their cargo back to Ynys Môn. They would make landfall in the early hours of the morning where local farmers had carts at the ready to unload and transport the goods before the customs agents arrived. The farmers also acted as lookouts. Many of the smuggled wares were moved off Ynys Môn and into mainland Wales but at least some of the good stuff was consumed here. One account written in 1760 noted the most popular drink on the island was a 'todi' – a sweetened mixture of brandy and water.

Enforcement was patchy, no doubt because even the local authorities had little appetite for preventing the illicit trade. The diaries of Lord Bulkeley, Sheriff of Anglesey, show that he was quite happy to purchase contraband brandy and claret while bemoaning an Act of Parliament that slapped high tariffs on soap and candles.

Without any top-level appetite for enforcement, it's no great surprise that the customs agents themselves weren't motivated to pursue the smugglers. One group that operated further down the coast were particularly fearsome. Not just because the men were thuggish but because their wives were said to put curses on local people. They were known as the Witches of Llanddona.

According to Ynys Môn folklore, a mysterious boat washed ashore in Red Wharf Bay, about twelve miles south of Amlwch. It had no rudder or oars and it was full of desperate men and women half dead with hunger and thirst.

The locals were convinced these new arrivals were criminals and prepared to drive the boat back into the sea. But, as they did so, the women in the boat commanded a spring of pure water to burst out of the sands. The locals, dumbstruck by the act and fearful of this magic, allowed the migrants to stay and build homes on the cliffside below the village.

The newcomers soon revealed their true nature. The men were ruthless smugglers and wore distinctive red neckties. They stole from the villagers and, when customs officers got too close, would release swarms of deadly black flies that would blind their opponents. The women were even more feared because of their witchcraft – they put spells and curses on farm animals and property and charged the locals money to remove them. No one knows quite how many of these witches lived in Llanddona but two, Bella Fawr (Big Bella) and Siani Bwt (Short Betty), are remembered to this day.

Who were the Witches of Llanddona? Some tellings of the story say they had Irish accents. In the seventeenth century, there was widespread fear among protestant communities of a Catholic Irish invasion following the aftermath of the English Civil War. This story might have been used to stoke fear among other local communities on Ynys Môn.

Or it could have been that the local Llanddona community, like many fishing communities whose livelihoods depended on the unpredictable rhythms of the sea and the weather, saw the arrival of outsiders as a bad omen or a sign of bad luck. Women were still being executed under British law for witchcraft up to 1725 and the practice wasn't made illegal until the Witchcraft Act was passed in 1735. A lingering suspicion about malevolent supernatural abilities would remain for many decades to come – especially in small rural and isolated villages. It's not hard to imagine this band of brigands being blamed for a failed fishing catch and bad weather, even in an age when the smartest minds were starting to analyse and understand the sea through the lens of mathematics and science. According to the 1909 book, the *Folk-lore and Folk-stories of Wales*, September gales were attributed to witches and were often called 'the witches' frolics', even in the early twentieth century.

Ahead of me lies Traeth Dulas, the wide, deep estuary of the Afon Goch. The map shows the Coast Path hugging and wrapping around the sides of the estuary. But from what I can see, there is a clear short cut across the *morfa* and sand banks. It will save me at least two hours of walking. I check the tides on my phone – there looks to be plenty of time before the sea comes rushing back into the bay – so I leave the path and step out confidently onto the salt marsh.

I pick my way through the shallow puddles in the sand and stride towards the middle of the bay. A sole cormorant wades in a shallow estuary stream off in the distance. A flock of sheep nibbles the long grass up on the hill. They are my only company on this beautiful stretch of sand.

And that should have alerted me to the folly of this short cut. Within minutes I encounter the main obstacle to success – a wide channel of water flowing out to sea. The other side of the bank looks tantalising close and almost achievable to ford. But, for once, common sense gets the better of me. I don't want soaking wet feet and I definitely don't want to get stranded on a sand bank when the tide returns. Defeated, I turn back and take the long way around.

Even by Ynys Môn's standards, this part of the island feels isolated. So it's even more remarkable that not one but two great nautical mapping minds hail from here, both of whom could have told me not to attempt the last short cut.

The first was William Jones, a mathematics prodigy who grew up in the parish of Llanfihangel Tre'r Beirdd, a few miles inland from where I'm walking.

Jones's genius was spotted at an early age and his education was funded by the local landowner Lord Bulkeley. Not surprisingly for a boy brought up near the Ynys Môn coast, Jones felt an affinity for the sea and one of his first jobs was teaching mathematics aboard British Navy ships. He was fascinated by the potential for mathematics to aid navigation (echoing his compatriot Robert Recorde more than a century before). In 1702, and now based in London, Jones published *A New Compendium of the Whole Art of Practical Navigation* where he demonstrated new mathematical methods of calculating a ship's position at sea.

That's not the reason Jones is famous today though. In 1706, Jones came up with the π symbol to describe Pi – the ratio of the circumference of a circle to its diameter. His simplification would take a concept that had been identified as far back as Babylonian times and mainstream its application. While the mathematical value of Pi is an irrational number (3.14… and then perhaps as much as 2.7 trillion other digits), that ratio is always the same value no matter the size of the circle it is describing. Pi's explanation of spherical systems means that it appears in many maths and physics formulas including in meteorology (it plays a key role in our understanding of rain drops) and astronomy. Today, we use Pi to understand atmospheric patterns and the movement of the winds – even fluctuations in the jet stream.

I reach the inland neck of Traeth Dulas, where the Coast Path turns back on itself and heads down the south side of the bay. Here, I pass an old farmhouse called Pentre Eirianell. This once was home to Lewis Morris, the second great nautical mind, and part of a famous local family known as Morrisiaid Môn (The Morrises of Anglesey). Over the course of his lifetime, Morris's passion for understanding and mapping the coast of Wales would transform how sailors navigated its waters.

Morris had showed an interest in coastal mapping from an early age; one of his first jobs was to map the holdings of another prominent local landowner called Owen Meyrick. In 1729, he was appointed as 'Searcher and Customs Officer' for the northern coast of Ynys Môn and it was here that he would likely have heard tales from travelling seafarers about the dangers posed by this section of the north Wales coast and the failings of the local charts that existed at the time. It had only been fifty years since the royal yacht *Mary* had run aground and been wrecked on the treacherous Skerries rocks off the island's north coast.

Morris had no formal training as a hydrographer but he sensed an opportunity. At the time, the lack of a road infrastructure made travelling overland in Wales both arduous and time-consuming. It was much quicker to carry goods and passengers by ship but navigating these mostly uncharted or badly mapped waters was risky.

In 1734, Morris proposed a project to map the entire Welsh coast but neither the British Admiralty nor the Customs Commissioners were keen to fund it. Undeterred, Morris hired his own boat and began work. Ultimately, Meyrick, the landowner who he'd first worked for, helped sway opinion with the Admiralty to support Lewis's effort.

Lewis had originally intended to keep his first drafts private but the Admiralty wanted the work shared because coastal shipping was on the rise. So, in 1748, he published his work – a small volume of twenty different coastal charts showing harbours and coves that could offer shelter in stormy weather. All were a major improvement on earlier charts and provided local information on tidal streams, hazards and safe anchorages.

Nothing of this detail and precision had been done before and Lewis's work quickly grabbed the attention of Britain's top mariners. More than 1,200 people subscribed to the first printing of the charts including Lord Nelson. Today, these works sit in the National Library of Wales and Lewis is considered one of the most eminent of British cartographers.

A very muddy hour or so after my aborted short cut, I arrive at Traeth Lligwy, a stunning long sandy beach just a couple of miles from the village of Moelfre. I cross a wooden footbridge over Nant y Perfedd, its pale blue freshwater flowing leisurely towards the bay. It is a perfect soft lit afternoon. The low sun brings out the pale green of the hillsides while the bracken appears more tan than red. The shallow cliffs cast a shadow over the sands and the sea has a purple sheen to it.

I continue past a solitary cottage named Moryn that sits at the southern edge of the bay. I've rarely felt as relaxed and at peace as I do on this isolated stretch of Ynys Môn. Yet, despite its tranquillity this particular afternoon, I'm actually looking out at the site of the worst maritime disaster ever to befall Welsh waters.

It was here, on the night of 25 October 1859, that the *Royal Charter* star clipper ran aground on the rocks with some 370 passengers and a crew of 112 on board. It was on the final leg of a two-month journey from Melbourne, Australia to its home port of Liverpool. Many of the passengers were gold miners who had struck lucky, so the boat was laden with their treasure.

All had been going well until the ship sailed past the Trwyn Eilian lighthouse. It was then that it was hit by a hurricane force twelve storm that drove it south and towards the coast. The *Royal Charter* dropped anchor but such was the force of the storm that the chain snapped. Within hours it had run aground first on a sandbank and then, catastrophically, it crashed into the razor-sharp jagged rocks at Porth Helaeth just north of the village of Moelfre – where I am right now.

Some 459 people perished that night including all the women and children. A number of them were swept overboard while waiting to be rescued so the rest were told to wait below deck. The ship broke into pieces before a rescue could be attempted. All told, across the United Kingdom, the storm of 1859 took 800 lives and 133 ships, with ninety more badly damaged.

News of the tragedy spread rapidly and within days the media had arrived in this quiet part of north Wales. Some of the sensationalist press ran spurious stories that local villagers had plundered the bodies of the dead, blaming the greed of 'Cambro-British thieves'. Later, another young journalist named Charles Dickens visited to write a more considered account. In a chapter titled 'Shipwreck' from his book *The Uncommercial Traveller*, he described how the local clergyman, Revd Stephen Roose Hughes, witnessed the *Royal Charter* as it ran aground and discovered: 'the sea mercilessly beating over a great broken ship.'

> They saw the ship's life-boat put off from one of the heaps of wreck; and first, there were three men in her, and in a moment she capsized, and there were but two; and again, she was struck by a vast mass of water, and there was but one; and again, she was thrown bottom upward, and that one, with his arm struck through the broken planks and waving as if for the help that could never reach him, went down into the deep.

The wreck of the *Royal Charter* also got the media speculating about why there had been no warning of such a destructive storm. These conversations got the attention of Admiral Robert Fitzroy. He had gained fame as captain of HMS *Beagle*, the ship that had taken naturalist Charles Darwin around the world on a voyage that helped form his ideas about evolution. While on board Fitzroy collected a set of astrological observations that would develop into a set of reliable longitudinal readings for the entire globe.

Fitzroy, now retired from the Navy, had recently founded the fledgling Meteorological Office and devoted his time to studying weather patterns around the British Isles. He decided to look deeper for meteorological clues leading up to what was being called the *Royal Charter* storm. By researching low barometric trends in the days before the storm, as well as severe weather reports from the Finisterre region of the Bay of Biscay, Fitzroy was able to produce charts that demonstrated how the storm could have been predicted. Having proved his model worked, he then proposed the creation of a national storm warning system.

Fitzroy faced a great deal of opposition and ridicule both from fellow scientists and politicians to what he referred to as 'forecasting'. Despite the many breakthroughs in science and engineering during the Victorian era, there persisted a sense of incredulity that human beings could predict something as God-given as the weather. Darwin and Alfred Russel Wallace's theories of evolution were only just gaining acceptance while only a few decades had passed since the pre-eminent geologist, William Buckland, had insisted that human existence only began after Noah's Great Flood.

Indeed, when one Member of Parliament had suggested in the House of Commons that recent advances in scientific theory might soon allow them to know the weather in London 'twenty-four hours beforehand', his fellow MPs roared with derisive laughter.

Despite the hostility he faced, Fitzroy persevered with his forecasting system, using the new copper-powered electric telegraph communication system to gather real-time weather data from the coasts at his London office. If Fitzroy thought a storm was gathering, he would telegraph a port where a drum was raised in the harbour indicating the direction of the wind. He described his work as 'a race to warn the outpost before the gale reaches them'.

Fitzroy's forecasting wasn't, of course, perfect and for every success he showed, his mistakes were vilified by his detractors. Over time, the campaign to undermine Fitzroy's innovative work took such a toll on his mental health that he took his own life. Ultimately though, his scientific approach to listening to the sea, understanding weather patterns and anticipating how they will change prevailed.

The organisation he founded became the Met Office that we know today. His storm warning service grew over time to become the Shipping Forecast – the longest running national forecasting service in the world. And, today, the Met Office's Hadley Centre for Climate Sciences and Services is one of the leading scientific authorities on climate change.

It's just a short walk into the village of Moelfre but it gives me a chance to reflect on this spectacular piece of coastline, albeit with a Jekyll and Hyde ability to transform from calm to chaotic.

Seascape

Since leaving Amlwch, I'd discovered how smuggling had flourished on this northern coast of Ynys Môn thanks to its proximity to the Isle of Man and its rugged, craggy hideaway bays that allowed contraband to be landed undetected by customs officials. For over a century, smugglers and law enforcement had sought to outwit each other through their navigation skills and local knowledge of this coastline. It was this coast that provided William Jones and Lewis Morris with their first education in navigation (the latter while working as a customs officer). And it was also Admiral Fitzroy's knowledge of navigating the coast that inspired him to push the boundaries of weather science. Like Lewis Morris, he had started his nautical career working with the Customs service patrolling the coast of Cornwall for smuggling.

I pass the stone memorial to the shipwreck victims that sits on a small hill above the low cliffs. A lone gull is perched on top casting a gaze towards the horizon. This part of the Welsh coast can be cruel and many lives have been lost to shipwrecks over the years. But it has also helped shape our skills of how to survive at sea and how to understand the weather systems that influence our daily lives. That is the enduring legacy of the *Royal Charter*.

14

The Winds of Change

Walks from Deganwy to Rhos on Sea and Rhyl to Talacre Beach

The view from Deganwy waterfront, at the mouth of the Afon Conwy Estuary, is one of those gems you'd prefer not to tell anyone about – so tranquil is its setting even though it sits just a couple of miles from Llandudno, Wales's biggest tourist town.

Here, the bay opens out towards Llanfairfechan and Ynys Môn beyond. Down the estuary lies Conwy Castle – another of Master Builder James of St George's 'ring of iron' defences – and spanning the river next to the castle are two more examples of the Thomas Telford/Robert Stephenson engineering double act.

In 1826, the same year that the Menai Suspension Bridge opened, Telford also completed work on a similar design here in Conwy. His goal was to design a bridge that honoured and mirrored the medieval design of the castle. However, to anchor the suspension cables, he had to demolish part of the old castle which, in hindsight, seems a strange homage. Then, in 1849, Stephenson finished construction of a new tubular-design railway bridge – just a few months before he would complete the Britannia Bridge. His bridge façade also tried to hark back to medieval times – albeit in a *Monty Python and the Holy Grail* type of way. But at least Stephenson didn't need to butcher the castle to honour it.

My route today will take me from Deganwy up to Y Gogarth (the Great Orme) and then from Llandudno to Rhos on Sea – about a thirteen-mile journey in total. Y Gogarth is a nearly two-mile-long slab of limestone headland that juts out into the sea just to the west of Llandudno. Its English name is said to come from the Norse word for dragon or viper. Folklore maintains that Viking sailors gave it that name even though most of the local Welsh-speaking population were quite happy calling it Y Gogarth until Llandudno was developed as a seaside destination for English tourists in the nineteenth century. That's when the name Great Orme became popular. To my eyes, when you look at the headland on the map, it more resembles a misshapen bottle opener. But I doubt the Vikings had much use for them.

It would have been easy to avoid the climb and cut inland to Llandudno Promenade at West Shore Beach but, sometimes, you have to embrace the challenge of going up a big rock – just to say you've done it.

The Coast Path follows the beach north up the headland weaving through a set of dunes that lie on the fringes of Maesdu Golf Club. A sign on the path explains the complexities of the system: the embryo dunes forming on the beach near the sea, the mobile dunes at the top of the beach and the fixed dunes that stand on higher ground further inland.

At West Shore Beach I follow Marine Drive as it loops around Y Gogarth. It started life as a footpath back in 1858. Later, it was turned into a road for carriages. Today, it's a one-way toll road for cars to reach the summit complex of the Orme. My route to the top is more direct. I leave Marine Drive and follow a public footpath that switchbacks up the side of the rock until it arrives at a long stone perimeter wall of a National Trust farm. In the distance, close to the summit, a farmer is herding sheep while riding his red all-terrain vehicle. Next to him, his sheepdog sprints back and forth, jumping from one boulder to another on the rocky green terrain.

The Vikings may get the credit for iconising this grand headland, but Y Gogarth was famous throughout Europe at least 2,500 years before. That was during the Bronze Age when it boasted the largest copper mines in all Western Europe and traders travelled across the Western Sea Routes and beyond to purchase copper ore. When heated with another ore, such as tin, zinc or magnesium, it combined to create the durable alloy that gave the epoch its name and transformed human life all around the world.

Given the importance of Y Gogarth through the ages, it's not surprising that it boasts its own mythic tale of a lost kingdom. This one concerns Tyno Helig, a sixth-century Welsh kingdom ruled by Prince Helig ap Glannawg and located on a low coastal plain to the west of Y Gogarth. The legend says that Helig had a daughter, Gwendud, who was both beautiful and cruel. She was expected to marry a nobleman called Tathal from Eryri (Snowdonia) but refused to do so unless he obtained a golden collar that would convey him high status. Tathal murdered a Scottish chieftain to steal his gold collar so he and Gwendud could marry. But on their wedding night, the ghost of the murdered chieftain appeared and set a curse on the couple. Soon after Tyno Helig was swallowed by the sea.

I follow the wall out towards the tip of the headland and pause to take in the view. To the west I can just see Ynys Seiriol and, behind it, Penmon Point on Ynys Môn. Ynys Seiriol provided solitude and sanctuary for its namesake saint back in the fifth century (the Vikings named it Priestholm – apparently having exhausted their creative juices coming up with Great Orme). Its English name is Puffin Island because, up until the late nineteenth century, it was a haven for breeding puffins. Then humans introduced the brown rat to the island and this invasive species decimated the puffin population. Today, the colony is slowly recovering after a rat eradication programme was implemented back in 1978.

To the east, the long and almost linear coastline of this part of north Wales spreads out into the distance. About five miles offshore stand the North Hoyle, Rhyl Flats and Gwynt y Môr windfarms – their turbines gently turning and pirouetting like giant ballet dancers. Together, these three major renewable energy sites generate more than 720 MW of electricity each year – enough to power around 500,000 homes. That capacity will increase so that it can power another 500,000 homes when the Awel y Môr extension to the Gwynt y Môr project, nine miles to the north of Y Gogarth, is up and running.

Sea-based renewable energy has the potential to transform modern Wales just as profoundly as this headland's ancient copper mines did back in the Bronze Age. Two-thirds of Wales is surrounded by the sea. Harnessing the energy already locked into the tidal currents and the wind to generate electricity could liberate the nation from its reliance on fossil fuels and, with the help of advancements in William Robert Grove's battery technologies, establish Wales as a net exporter of renewable energy.

The world has understood the power of wind energy ever since humans learned to sail. But the first known example of using wind turbines to produce electricity happened in Scotland back in 1887. That year, Professor James Blyth installed a ten-metre-high cloth-sailed wind turbine in the garden of his holiday cottage and used it to power the internal lighting in the cottage. It generated more than enough electricity so Blyth offered the surplus to the local community to light the main street. They declined, saying electricity was 'the work of the devil'.

In the twentieth century, despite being surrounded by potent weather systems that roll in from the Atlantic and North Sea, the United Kingdom showed only a tepid interest in wind power even as other European nations like Denmark invested heavily in diversifying their power grid. Part of the reason was out of concern that wind power wasn't dependable enough compared to North Sea oil, coal and nuclear (an argument that those industries invested heavily in to promote).

In the late 1970s, however, following the decision by Middle East oil producers to demonstrate their geopolitical influence by reducing crude oil supplies and hence causing shortages and wild price hikes, the UK government began exploring wind power technology to spread its energy risk.

They established the Carmarthen Bay Wind Energy Demonstration Centre on the Llwchwr Estuary, just outside of Burry Port, where Jeff and I had walked a few months before and discovered the story of Amelia Earhart and of the many shipwrecks on the sands.

From 1984 to 1991, a number of different horizontal and vertical axis turbines were prototyped on the site of an old coal fired power station owned and operated by the Central Electricity Generating Board (CEGB). By far the strangest was one nicknamed the 'magic mushroom'. This vertical axis wind turbine (VAWT) was six metres in diameter and had five vertical blades that were spring-operated to avoid rotor over-speeding in high winds. After twenty-one months of testing, it was retired as 'non-cost effective' but its novelty value (along with the other prototypes) drew curious wind 'tourists' from far and wide. Up to 16,000 people each year visited the demonstration site while it was in operation.

If you walk along the beach outside Burry Port, you can still see the ruined octagonal base station of one of the turbines – a Musgrove VAWT.

No one doubts how the sea can help the energy mix anymore. Off the coast of west Wales, licences were recently granted to build a series of floating windfarms that could create more than 20,000 jobs and generate four GW of electricity. Just a few miles away from here, off the coast of Ynys Môn, plans are afoot to build one of the largest tidal stream energy sites in the world, covering thirteen square miles of the seabed. Called Morlais, this marine project aims to generate kinetic energy from tidal currents. And right here, in the waters below Y Gogarth, developers dream of a grand tidal lagoon stretching from Llandudno to Prestatyn nineteen miles down the coast. The project would cost £7 billion to build but could potentially power more than a million homes and create another 20,000 jobs.

The tidal lagoon is the latest in a long list of proposed projects to generate power through lagoons and barrages. The most notable one is the ongoing campaign to build a barrage across the Hafren Estuary, where I started this journey, as a way of harnessing the power in its tidal reach.

The very first plan was hatched in 1849 by Thomas Fulljames, an architect and the county surveyor in Gloucestershire. He wanted to build a barrage from Beachley to Aust, close to the site of the first Severn Bridge. In recent decades, a number of proposals have been floated to generate renewable energy through a much wider tidal range (as barrages are known in the infrastructure sector). Academic estimates suggest a barrage across the Hafren could generate eight GW of electricity each year. That's enough to power the whole of Wales.

Each time though, the proposals for the Hafren have faltered because politicians struggle to believe that they are economically feasible and also because building the barrage would fundamentally alter the physical landscape of the Gwent Levels – threatening the rich biodiversity of the UK's 'Amazon Rainforest' by flooding local nesting grounds and wildlife preserves like the nearby Goldcliff Lagoons.

In Swansea, entrepreneurs have been lobbying to get support for a £1.7 billion tidal lagoon that would generate electricity from the bay – which also has a strong tidal reach. Versions of this plan have been floated and amended for the past twenty years. Each time, funding issues and a lack of UK government support have stalled any progress.

Today, the threat of climate-driven flooding is even greater, as is the need to ditch fossil fuels and embrace renewable energy. At the same time, the cost of building tidal infrastructure is cheaper than before and new technology has made the operation of these projects more efficient. Crucially, political and public opinion is shifting rapidly in favour of renewable energy and many of the arguments that once derailed these mega projects appear to be dissipating.

If Wales does have to defend its coast against the energy thrown at it by the sea, it might as well benefit from the billions in pounds of revenues that will come from harnessing that energy through renewable ventures.

Except, at present, it won't because the Welsh government doesn't own or control its own coastal seabed. The Crown Estate, owned by the British royal family, does. It is independently run but exists to generate money for the UK Treasury. In 2023 alone, The Crown Estate signed lease agreements for six offshore wind projects around the UK (including Wales) that will provide it with about £1 billion a year in revenue.

Given that Wales, as part of its devolved powers within the United Kingdom, has responsibility for its environment and at least some of its energy budget, it would seem to be logical that Crown Estate revenues collected around Welsh waters – both for the leasing rights to build windfarms and the annual rents – should be shared with the Welsh rather than the UK government. There is, after all, a proven model for how national governments can best use the fruits of their own natural resources. In Norway, the Sovereign Wealth Fund – established in the 1970s to invest its North Sea oil rents for the benefit of greater society – now exceeds £1.1 trillion. That is fifty times the entire Welsh government budget, most of which is handed down from the UK government in Westminster in the form of a grant.

To understand just what a difference those billions in revenue from a devolved Crown Estate could make to Wales's climate change protection efforts, you just need to walk down from Y Gogarth and onto Llandudno Promenade.

Llandudno is one of Wales's grandest Victorian era seaside resorts, its centrepiece being the two-mile-long promenade that is lined with hotels, and curves elegantly around the bay. The promenade offers direct access to the beach which has attracted tourists for more than 150 years.

It was the vision of the Mostyns, an aristocratic dynasty with ties to north Wales dating back to the twelfth century. In 1849 a Liverpool architect called Owen Williams presented Lord Mostyn with a proposal to develop the marshland he owned between Conwy Estuary and Llandudno Bay as a holiday resort. Over the next thirty years that marshland would be transformed into what became known as 'the Queen of Welsh Watering Places'. In time, nearly all the north Wales coastline would be developed for tourism. Today, the resorts of Llandudno, Rhyl and Prestatyn bring in nearly £2.5 billion each year in tourism income.

North Wales, however, is also one of Wales's most flood-threatened coastlines. One recent study found that Llandudno was the fifth most likely UK town to flood – no great surprise when you consider that most of the town sits at or below sea level. Natural Resources Wales Flood Development Advice Map has identified the entire north Wales coast as a serious risk. Tourism aside, this is particularly significant because north Wales's two main transportation corridors – the North Wales Expressway and the coastal railway, both linking the UK to Ireland – run through the flood risk zone.

Flood defences already protect twenty-eight per cent of the Welsh coast and some £8 billion of assets. The Flood and Coastal Erosion Management Programme has allocated £75 million to bolster those barriers across Wales including here in Llandudno and also in Rhyl, where a £27.5 million scheme includes the installation of rock armour to raise the existing sea wall and protect up to 1,650 homes.

How though do you effectively build sea defences when you also need to keep the coastline attractive for tourists? That's one of the major dilemmas facing Llandudno.

Back in 2014, some 50,000 tonnes of stones were dumped on its North Shore Beach to protect against erosion and flooding. So far, the defences have proved effective but many local people complain that the stones have destroyed the beach's natural appeal and are hurting tourism. They say they want their sand beach back – a demand that is growing now that the locals have seen how the beach at Rhos on Sea, a resort just down the coast, is being shored up with one million tonnes of sand – some of it pumped up from the seabed.

The work at Rhos on Sea is a form of sandscaping, a Dutch concept that can be highly effective as a sea defence as the sand increases the height and width of the beaches and so helps dissipate and absorb the force of the ocean. That said, it only works in certain environments – long unbroken shorelines are needed so that the sand spreads effectively. It can be counter-productive if added to protected habitats like existing dune complexes.

There's also another issue – sandscaping costs a lot more money than stone and rock defences. The sand project at Rhos on Sea cost £14 million and the estimated cost of replacing the stones at Llandudno with sand would be £24 million. The current rock protection scheme only costs £7 million.

How to pay for the amount of coastal protection needed will be a major political debate in the coming decades. Recent estimates suggest that the number of properties facing a significant risk of flooding will increase by more than seventy-five per cent by 2035 if Welsh government investment in managing flood risk continues at its present rate. Yet, based on the current demands for public funding, the Welsh government will struggle to afford the new defences – stone or sand.

Seascape

I reach Rhiwledyn (Little Orme), the younger sibling headland to Y Gogarth at the other end of Llandudno Bay. From this vantage point, this rock seems perfectly formed. It is only when I get to the other side that the real story reveals itself. In the nineteenth century this was a limestone quarry. Much of the rock has been blasted away but only this side; Lord Mostyn banned any quarrying on the Llandudno side so as not to ruin the view from his new tourist town.

It's particularly busy on the path today. A group of tourists has walked up here in the hope of spotting the seals who come to Porth Dyniewaid to nurture their pups. I press on, walking down the coast road that runs through Penrhyn. Here, I see more aggressive defences being laid along the north Wales coast. Row upon row of grey stone boulders – each about two metres square – have been deposited on the beach in front of the existing concrete sea wall. It looks like Penrhyn is ready to go to war with the sea – a show of strength and protection but one that surely ruins the main reason for going to the beach. I'm reminded of the response from a US general during the Vietnam War when asked why it was necessary to carpet bomb the town of Ben Tre. 'It became necessary to destroy the town to save it,' he said.

Directly opposite the new beach defences lies a small stream that runs alongside the local golf club. There's a plaque on a nearby bench that reads:

> Prince Madoc sailed from here, Aber Kerrik Gwynan 1170 AD and landed in Mobile, Alabama with his ships *Gorn Gwynant* and *Pedr Sant*.

Madoc is one of the most aspirational, enduring and malleable of all Welsh legends. The first mention of his voyage of discovery to America appears in a fifteenth-century poem by Meredudd ap Rhys where he compares his attempts at fishing to the exploits of

The Winds of Change

Madog ab Owain Gwynedd 'who sought nor lands nor flocks nor herds save in the vasty deep'.

The Madoc story was next championed during the Elizabethan age as a way of bolstering the English crown's claims in the New World over those of Spain. During the eighteenth century, antiquarians like Theophilus Evans and, later, the arch embellisher Iolo Morganwg promulgated the tale of Madoc and the existence of a Welsh-speaking Native American nation living somewhere along the Missouri River to emphasise the importance of Welsh heritage in British identity.

Whatever the truth about this seafaring Welsh prince, one thing is clear. If he tried to set sail from here now, his ships wouldn't make it past the giant sea defences.

There is one final walk to accomplish – an eight-mile straight shot along the beachfront from Rhyl to the Point of Ayr lighthouse at Talacre. Summer truly has faded into autumn and a low hazy sun offers brightness but little warmth as I walk past the Gorsedd standing stones on the long promenade. They were laid there in 1903 to commemorate Rhyl's hosting of the National Eisteddfod, the annual Welsh language cultural festival that was first staged back in 1176, not long after Madoc set sail.

Construction workers are operating cranes and heavy dozers to put more defences in place down on the beach. A short man in sunglasses and a green polo shirt stands watching the work. I ask him how long this coastal protection project has been going on.

'About a year. But it seems like they just move one set of big rocks from one place to another. They are spinning it out if you ask me,' he says rubbing his fingers together to simulate money grabbing.

As with Llandudno, some in the local Rhyl community consider the sea defences to be more of a hindrance to the tourism trade than a help to the town as a whole.

'Residents don't ever recall Rhyl High Street or the actual prom, apart from large waves, actually being flooded, and they are asking when was it last flooded?' complained one councillor at a public meeting, adding: 'It's also been noted that they think, because of the length (of time) of construction, it will be the death knell of Rhyl.'

There are plenty of people not just in Wales but all over the world who have the same attitude – we've never been flooded before so why should we worry now?

At the edge of town, more major and urgent work is taking place by Rhyl Golf Club in the hope of protecting over 2,000 homes. I pass Ffrith beach where tall dunes shield a sprawling caravan park from the sea. Kids amuse themselves by rolling down the dunes and digging sandcastles on the beach despite the autumn chill. Looking back down the coast, I can see Y Gogarth in the far distance. Those Vikings were right. It really does look like a sea serpent.

There's a cut through off the dunes into a housing estate on the edge of Prestatyn. It sits no more than 100 metres from the beach. I take the detour to see the houses because I want to get a sense of just how much protection the sand dunes might offer. From the edge of the development, the houses look safe from harm and yet it's the real vulnerability of these streets that is causing the national and local government to spend £26 million on better sea defences here.

Prestatyn, Rhyl and Llandudno can at least take heart that they are not alone in preparing now for something they can't envisage, but that we know is coming. Coastal cities and towns the world over are taking steps to build resilience in the best way for their own specific location.

Take Tokyo, where super levees are being built to protect the low-lying parts of the city. They have a much wider footprint and a flatter, more gradual slope than traditional levees. It takes a lot of land to build them but they repay the investment both through greater flood protection and also by providing a raised land area that can be developed for housing, schools and natural habitats.

Boston in the US is another coastal city that is tapping into its natural strengths by restoring marshes and creating waterfront parks to increase coastal flood protection and boost access to green space. In the Makoko slum neighbourhood on Lagos Lagoon in Nigeria, meanwhile, many homes are built on stilts and canoes are used as taxis. There is even a floating school that sits on top of empty barrels.

Jeff Goodell, author of *The Water Will Come*, has documented how some of the US's biggest cities are looking to repel the oceans. He describes how New Orleans has raised its defensive dykes, how New York City plans to build an enormous wall around Lower Manhattan and how Miami passed a $400 million 'Miami Forever' bond, a portion of which will fund sea walls to help protect the city.

Not surprisingly, one of the most successful flood protection infrastructure projects is in the Netherlands where the Maeslant barrier protects the port and city of Rotterdam by opening and closing in response to water level predictions calculated by a centralised computer system.

Here's the problem, though, with isolated sea defences: as we know, water is just liquid energy and if that energy is blocked in one area it flows to the next place of least resistance. Having a wall or barrage simply deflects the power of the water to other less resilient places. So communities and authorities will have to work in concert to make sure one successful form of coastal protection doesn't create problems for others. Also, the type of coastal defence can cause additional problems. Rock armour already feels like an industrial solution to a natural challenge but it can be even more detrimental if the type of rock chosen doesn't complement the biodiversity of the coast where it is being used. One of the major challenges in designing effective coastal barriers is to reduce wave action and erosion while also creating the right physical conditions to mimic the local habitat for marine creatures. Choosing the correct type of rock – often calcium rich limestone with rough surfaces – helps ecosystems colonise the structures, boost biodiversity and strengthen protection from waves.

In some areas of the coast, architects are designing artificial defence systems that mimic the protection offered by nature. In Blackpool, about forty-five miles north of Rhyl, the sea wall has been designed to deflect the sea in the same way that its sand dunes once did before (and this won't surprise you) Victorian engineers stripped them away to build a promenade.

I arrive at Prestatyn beachfront. The promenade and the long shallow beach are buzzing with weekend visitors. The gulls are eyeing up bags of chips, sandwiches and even ice creams in the hands of unsuspecting targets. In the distance, a storm is gathering over the Clwydian mountain range while, out to sea, the wind turbines turn at a steady if ponderous pace.

I stop for a break and think back to all the new sea defences I've seen over the last few days. Llandudno, Penrhyn, Rhos on Sea, Pensarn, Towyn, Rhyl and now Prestatyn. It all feels a little medieval, a rear-guard effort to protect our consistent but often misguided longing to live by the sea.

I think back to Robert Fitzroy and how he put his faith in data – charts back then – to help forecast what the weather might bring. Engineers like Sir John Hawkshaw of the Severn Tunnel, Thomas Telford and Robert Stephenson all believed in the power of construction to tame or overcome the challenges of the sea. Fitzroy chose a different approach. He wanted to learn from the sea to understand where and how it would demonstrate its power. Perhaps having been a ship's captain made him appreciate nature's force more than most. In that respect he was following in the footsteps of Robert Recorde and William Jones – both of whom used mathematics to interpret the stars – and even Daniel Defoe whose journalistic reporting offered new insights on how storms impact a whole country.

Despite all the brute force being applied to create obstacles along the coast, we, as a Welsh and global society, are going to have to think differently about our relationship with the sea. The more we understand how it is changing, the better prepared we will be to make the right decisions about what we protect and what we surrender in the decades to come.

Today, our ability to collect and process data is greater than ever before. And with the growth of artificial intelligence and data-led decision making, we have the tools and insight to analyse and predict where and how the sea and storm risks will come. That will deliver great benefits in anticipating the coming climate risks. But with knowledge comes responsibilities and that will be bad news for millions of people whose houses and towns were built without the insight we now have.

The global insurance industry already uses AI and predictive intelligence to assess property risks. Swiss Re, the reinsurance giant, projects that climate-related risks could add an additional $183 billion to property insurance costs by 2040. We've already seen how insurers are refusing to provide coverage for properties in vulnerable parts of the coast globally. As AI climate modelling improves, more and more people will discover that their coastal properties are essentially worthless.

How long will it be before the residents of Prestatyn (and other coastal residents all around Wales) are told their own properties have been deemed uninsurable – as the village of Fairbourne already has. When that happens, property values plummet and people are left with what economists call stranded assets. New people don't want to move into an area that is uninsurable and, frankly, local governments won't want to spend the money to protect those areas because they won't be receiving the tax revenue needed to make those investments.

Seascape

I leave Prestatyn and walk through the windswept Gronant Dunes Nature Reserve to Talacre Beach and the Point of Ayr. The path leads over the crest of the dunes and the tall anchoring grass whistles in the strong breeze while protecting a blanket of orchids growing in the sand. Groups of black-headed and white-bodied little terns peck around in the sands close to the sea. Within days they will depart these shores for their winter sojourn in West Africa.

The waves have an aggressive rush and whirl to them as I walk across the beach – like they are warming up for some major activity. The sand at the base of the red-topped Point of Ayr lighthouse has mixed with mud to form small depressions that are half full of sea water. Off in the distance, the white wind turbines seem almost in touching distance.

I sit at the top of the beach, where the hard sand meets the fine grains of the dune complex, and consider the journey I've been on. From my starting point in Chepstow I've covered nearly 400 miles – just under half of the total 870-mile-long Coast Path. It isn't a comprehensive examination of the Welsh coast but enough to experience its beauty and variety, to reflect on how it has changed over the centuries and how it will keep changing in the coming decades.

All along the Coast Path, I've discovered how the sea has shaped our lives in Wales and how we have tried to shape and tame it – constantly developing projects to reclaim salt marsh, to build causeways and barrages, and to put the sea to use for industry and for food. I've learned some of the many stories, tales and legends that have been passed down over the centuries to mythologise the sea – often imbuing it and the people who come under its spell with special skills and powers. I've always sensed the sea was a defining influence on my life. Now I can see just how much it is embedded in the consciousness of Wales as a whole.

I walk east from the lighthouse to the very tip of the sand banks – where the sea meets the fresh water flowing out of the

Dee Estuary. I can just make out the English coast of The Wirral and the town of Hoylake on the other side. I stop by a cluster of sandcastles built earlier in the day – elaborate fortifications that some small child (and enterprising parent no doubt) have put satisfying effort into planning.

This is one of the most ecologically important parts of the entire Welsh coast but also potentially one of the most vulnerable to sea level rise. At some point over the next few decades, the tide, which rises and falls twice a day over these sand banks, will probably stop retreating. How the sea's power changes this and the rest of Wales's coastline is a question we still don't fully know the answer to. But we do know that change – inevitable and irrevocable – is coming. I leave the sandcastles to crumble under the weight of the coming tide. Tomorrow, perhaps, another child will come here and build new ones. And the day after and the day after that. Until, that is, there may be no more beach to play on.

Selected Bibliography

Aldhouse-Green, Miranda, *Enchanted Wales: Myth and Magic in Welsh Storytelling* (Cardiff: Calon Books, 2023)
Badder, Delyth and Norman, Mark, *The Folklore of Wales: Ghosts* (Cardiff: Calon Books, 2023)
Barber, Chris, *Mysterious Wales* (Stroud: Amberley Publishing, 2016)
Beazley, Elisabeth *Madocks and the Wonder of Wales* (London: Faber & Faber, 1967)
Blackburn, Julia, *Time Song* (London: Vintage, 2020)
Bollard, John K., *The Mabinogi* (Llandysul: Gomer Press, 2006)
Boorman, David *The Brighton of Wales* (Swansea: Swansea Little Theatre Company, 1986)
Brominicks, Julie, *The Edge of Cymru* (Bridgend: Seren, 2022)
Bullough, Tom, *Sarn Helen* (London: Granta Books, 2023)
Crickhowell, Nicholas, *Westminster, Wales and Water* (Cardiff: University of Wales Press, 1999)
Cunliffe, Barry, *Facing the Ocean* (Oxford: Oxford University Press, 2001)
Czerski, Helen, *Blue Machine* (London: Torva, 2023)

Selected Bibliography

Davies, John, *The Making of Wales* (Stroud: Sutton Publishing, 1996)

Davies, John, *A History of Wales* (London: Penguin, 1997)

Driver, Toby, *The Hillforts of Iron Age Wales* (Eardisley: Logaston Press, 2023)

Finch, Peter, *Edging The City* (Bridgend: Seren, 2022)

Graves, Carwyn, *Welsh Food Stories* (Cardiff: Calon Books, 2022)

Graves, Carwyn, *Tir* (Cardiff: Calon Books, 2024)

Hughes, Geraint Wyn, *Secret Anglesey* (Stroud: Amberley Publishing, 2019)

Jones, David Ceri, *The Fire Divine* (London: IVP, 2015)

Loveluck-Edwards, Graham, *Legends and Folklore of Bridgend and the Vale* (Cardiff: Jelly Bean Books, 2020)

Meirion, Dafydd, *Welsh Pirates* (Aberystwyth, Y Lolfa, 2006)

Miles, K. G., and Towns, Jeff, *Bob Dylan and Dylan Thomas* (Carmarthen: McNidder & Grace, 2024)

Parker, Mike, *All the Wide Border* (Manchester: HarperNorth, 2023)

Pennick, Nigel, *The Celtic Saints* (New York: Sterling Publishing Co. Inc., 1997)

Rhys, Gruff, *American Interior* (London: Penguin, 2015)

Ritchie, Hannah, *Not the End of the World* (London: Chatto & Windus, 2024)

Royal Commission, *Wales and the Sea: 10,000 Years of Welsh Maritime History* (Aberystwyth: Y Lolfa, 2019)

Stephens, Meic, *A Cardiff Anthology* (Bridgend: Seren, 1996)

Thomas, David N., *Dylan Thomas* (Bridgend: Seren Books, 2000)

Williams, Gwyn A., *When was Wales?* (London: Penguin, 1991)

Acknowledgements

Walking for me is sometimes about going solo but often about tramping with friends. This book wouldn't have happened without the support and encouragement of my good friends Andy Bethell, Jeff Jones, Tim Powell, Alan Tapp and Geoff Jones who joined me on different parts of the journey. Even when they weren't walking, their enthusiasm for this project propelled me forward.

Other friends, Elizabeth Hansen and Alexis Barthelay, joined me for walks while Peter Kellam, Iestyn Humphreys, Ian Buckland and Martin McCabe kept me on my toes by asking smart questions about my journey and giving me pointers on new areas of interest to consider.

The whole team at Calon have been enormously supportive and nurturing – particularly my triumvirate of editors, Amy Feldman, Abbie Headon and Clare Grist Taylor, whose ability to edit out my flights of self-indulgence has surely arrived far too late in my writing career!

I'm also very grateful, once again, to be working with Neil Gower whose visual creativity both inspires me and makes my work look so much better.

I'd particularly like to thank my friends Aminah and Davy who

Acknowledgements

offered up their home as a bolthole for me to go and write when I most needed time alone.

As always, my appreciation goes to my agent, Sharon Bowers, for giving me the advice I need when I need it.

Finally, I would not have been able to write this book without the support of my wife Jowa, who is always there as a sounding board for my creative ideas and also to pick up the pieces of family life when I go down a rabbit hole of walking and writing.